AGRICULTURE COMMITTEE

ORGANIC FARMING

Volume I

Report and Proceedings of the Committee

Ordered by The House of Commons *to be printed*
17 January 2001

PUBLISHED BY AUTHORITY OF THE HOUSE OF COMMONS
LONDON – THE STATIONERY OFFICE

£9·70

HC 149-I

The Agriculture Committee is appointed to examine on behalf of the House of Commons the expenditure, administration and policy of the Ministry of Agriculture, Fisheries and Food (and any associated public bodies). Its constitution and powers are set out in House of Commons Standing Order No. 152.

The Committee has a maximum of eleven members, of whom the quorum for any formal proceedings is three. The members of the Committee are appointed by the House and unless discharged remain on the Committee until the next dissolution of Parliament. The present membership of the Committee is as follows:

Mr David Borrow (*Labour, South Ribble*)
Mr David Curry (*Conservative, Skipton and Ripon*)
Mr David Drew (*Labour, Stroud*)
Mr Alan Hurst (*Labour, Braintree*)
Mr Michael Jack (*Conservative, Fylde*)
Mr Paul Marsden (*Labour, Shrewsbury and Atcham*)
Mr Austin Mitchell (*Labour, Great Grimsby*)
Mr Lembit Öpik (*Liberal Democrat, Montgomeryshire*)
Mr Owen Paterson (*Conservative, North Shropshire*)
Mr Mark Todd (*Labour, South Derbyshire*)
Dr George Turner (*Labour, North West Norfolk*)

On 15 February 2000, the Committee elected *Mr David Curry* as its Chairman.[1]

The Committee has the power to require the submission of written evidence and documents, to examine witnesses, and to make Reports to the House. In the footnotes to this Report, references to oral evidence are indicated by 'Q' followed by the question number, references to the written evidence are indicated by 'Ev' followed by a page number.

The Committee may meet at any time (except when Parliament is prorogued or dissolved) and at any place within the United Kingdom. The Committee may meet concurrently with other committees or sub-committees established under Standing Order No. 152 for the purpose of deliberating, taking evidence or considering draft reports, and with the House's European Scrutiny Committee (or any of its sub-committees) and Environmental Audit Committee for the purpose of deliberating or taking evidence. The Committee may exchange documents and evidence with any of these committees, as well as with the House's Public Accounts and Deregulation Committees.

The Reports and evidence of the Committee are published by The Stationery Office by Order of the House. All publications of the Committee (including press notices) are on the internet at www.parliament.uk/commons/selcom/agrihome.htm. A list of Reports of the Committee in the present Parliament is at the end of this volume.

All correspondence should be addressed to the Clerk of the Agriculture Committee, Committee Office, 7 Millbank, London SW1P 3JA. The telephone number for general inquiries is 020 7219 3262; the Committee's e-mail address is: agricom@parliament.uk.

---

[1] On 16 July 1997, the Committee elected Mr Peter Luff as its Chairman. He was discharged on 21 February 2000.

TABLE OF CONTENTS

Page

SECOND REPORT

The Agriculture Committee has agreed to the following Report:-

ORGANIC FARMING

FOREWORD

1. **The demand-led expansion in organic production in the UK has brought great benefits in revitalising this sector at a time of great trouble in the rest of the agricultural industry. This is to be welcomed and we applaud the efforts of the organic movement in responding to this demand so swiftly. Inevitably, however, the sudden increase has led to problems in oversubscription of organic services, from Government grants to certification of farms and imported products. There is an argument over whether the Government should invest now to meet more of the current level of demand from domestic supplies or whether in the longer term this would do more harm than good by creating a sudden glut on the organic market, an agricultural equivalent of a boom and bust economic cycle. Indeed, there is a real question as to the extent to which the Government should be providing support at all when the market is so obviously strong. We believe that there is a strong case for caution. There are also fears that the growth in organic demand is leading to a loss of control by the industry over its traditional values and principles, as larger and more commercially-oriented farmers and the supermarkets become ever more dominant in the market. These difficulties can be resolved by the industry acknowledging the fears and by working towards better supplier relationships and stronger producer-controlled co-operatives.**

2. **There is clearly a strong consumer demand for organic products but we are very conscious that the consumer may attribute benefits to organic products which cannot be sustained in the present state of scientific knowledge and which cannot legally be claimed by producers. We have reservations about the claims made for organics and we believe that far more work needs to be done to establish a scientific basis for these claims. This would then sustain a rationale for the standards applied and, together with research into technical issues, could lead to great advances by organic farmers. It is vital that consumers get what they believe they are paying for, which is why we attach such importance to clear standards. It is also vital that the taxpayer gets what he or she is paying for, which is why we support an organic stewardship scheme under which Government grants would reward proven environmental benefits.**

3. **We end by stressing the need to see organic and conventional agriculture as interdependent. We wish to see the best techniques of both systems used to ensure the greatest benefits for farmers, consumers and the wider community. It is unlikely that organic farming can ever provide the amount of food needed for the whole country so conventional agriculture will continue to play a major part, making it all the more important that the systems work in tandem and learn from one another. Organic farming has much to offer in this partnership and we hope that it will continue to develop and expand in the UK. Organic farming is now a mature sector. Some of its apostles still proselytise with an almost religious fervour and, occasionally, a sectarian spirit. This helps nobody. The past perhaps belonged to messianics; the future belongs to marketing.**

I. INTRODUCTION

4. Organic farming is one of the few bright spots in the depressed picture of UK agriculture. The number of producers expressing interest in conversion has grown exponentially over the last few years in response to the strong consumer demand for organic products. The speed at which this market has developed has resulted in many sectors in a huge gap between demand and the ability of the UK industry to supply, with the happy consequence for organic farmers of stable prices above those achievable for conventional supplies and the less welcome consequence for the UK's balance of trade of a reliance on imports. Traditionally, organic farms have been seen as small family-run enterprises and indeed many such farms do still exist. However, within the organic sector, there are also farms on a scale similar to those found in conventional farming and it is possible that more will be established in order to satisfy the level of demand. It is important to recognise that generalisations about organic farming are meaningless and that the sharp distinction often made between organic and conventional production is far less clear in reality.

For example, it is as possible to have an intensively farmed organic business as it is to have an extensively farmed conventional one. Similarly, conventional farmers may adopt practices associated with organic farming without seeking a change in status, and very interesting results have been achieved through integrated crop management which combines beneficial natural processes with modern farming techniques. Conventional and organic farming are not two wholly distinct activities and practitioners in either sector may have more in common than they have at variance. Some farmers will produce in both markets to maximise returns and to diversify their interests. In this Report, we are concerned not with the promotion of organic farming as a public good in itself but with finding answers to a technical problem – how to increase UK production to meet an existing demand – and with the question of appropriate Government support for organic farming.

Reality v. perception

5. The term "organic" refers to a *process*, not the final *product*. The entitlement to label vegetables, meat or any other foodstuff as organic depends upon the way in which it was produced and the procedures involved in processing, rather than any intrinsic, testable quality in the food itself. Many claims have been made for organically produced foods, ranging from food quality, food safety, animal welfare, support for rural communities and fair trade, and benefits for the environment. **We have seen no evidence to enable us to state unequivocally that any of these claims are always and invariably true. All claims need to be properly evaluated in order to help consumers make their own judgements on the benefits of organic produce.** Indeed, the organic movement itself, in general, is careful not to assert such claims as provable. Doubts have been raised, for example, on aspects of animal welfare, and the spectacle of green beans flown in from Kenya with high energy consumption is contrary to the organic ideal of locally grown, seasonal produce. We recognise that, of the claims made, by far the strongest case is that organic farming is environmentally more beneficial than conventional farming, although even here some organic practices, such as the use of the highly toxic copper sulphate (to be phased out shortly), synthetic pyrethroid sheep dips or the killing of weeds by flame, are less environmentally-friendly than the equivalent conventional practices. It is a common perception that organic means pesticide- and chemical-free but in fact it simply means farming without *artificial* pesticides: those produced from natural chemicals may be used. In the same way, there is a significant list of non-organic processing aids which may be used in manufacturing organic products and a tolerance level of five per cent non-organic ingredients in processed products labelled as organic.

6. This is not to accuse the organic movement of misleading the public but it is perhaps true that the public has a perception of organic farming that is, at least partly, mythical. **We believe it important that the claims can be tested and verified in order that consumers know what they are really buying.** The statement from the Food Standards Agency (FSA) in August 2000 that it "considers that there is not enough information available at present to be able to say that organic foods are significantly different in terms of their safety and nutritional content to those produced by conventional farming" raised a furore, but illustrates the limits of claims which can be scientifically sustained.[1] Research to sustain or quantify the claimed benefits of organic farming is badly needed. However, it is clearly the case that some consumers in the UK wish to buy organic products and, whatever their reason for doing so, be it some dream of the perfect English countryside or fears over food safety, they have the right to do so as long as those products meet all legal safety standards.

The Lords Report

7. In July 1999, a year before our own inquiry began, the Select Committee on the European Communities of the House of Lords reported on *Organic Farming and the European Union*.[2] The report was impressive, covering all the main questions on organic farming, including discussion of its philosophy and principles. We have read the analysis and conclusions of their Lordships and have felt excused from examining certain areas in order to avoid duplication. But the year between the end of their inquiry and the beginning of ours has seen the organic market

---

[1]FSA Position Paper: Food Standards Agency View on Organic Foods, August 2000.
[2]16th Report of the Select Committee on the European Communities, Session 1998-99, HL Paper 93.

continue its phenomenal growth (made all the more striking by the continued fall in conventional prices); it has seen the implementation of European-wide livestock regulations in August 2000, closing one of the gaps in the standardisation of organic practice; finally, it has also seen the opening – and premature closure to new applicants – of the Government's organic farming scheme and the announcement of additional new funds for organic conversion through the Rural Development Regulation. These factors justify a new look at the sector.

Conduct of inquiry

8. We announced our inquiry in a press notice issued on 5 May 2000, calling for evidence on:

> the expansion of organic farming in all agricultural sectors; market trends and customer demand; the role of organic certification organisations; the setting of organic standards and tolerances; the role of farm assurance schemes; the availability and suitability of public and private assistance for organic conversion, including the role of trade associations, food processors, supermarkets and the Government; outlets and distribution systems for organic produce and retail pricing; the level of imports and exports of organic foods; international comparisons; and likely future developments in these areas.

In response, we received over 70 memoranda, a high total which reflects the extent to which this subject has caught the public attention. In addition, we held five oral evidence sessions, hearing from Professor Sir John Marsh CBE, Professor William McKelvey of the Scottish Agricultural College, Dr Nicolas Lampkin of the Organic Farming Centre for Wales, the National Farmers' Union of England and Wales, three organic farmers (Mr Nick Bradley, Mrs Joanna Comley and Mr Oliver Watson), Yeo Valley Organic Company Limited, Congelow Produce Ltd, J Sainsbury plc, Iceland Frozen Foods plc, the Soil Association, Organic Farmers and Growers Ltd, the United Kingdom Register of Organic Food Standards (UKROFS) and Mr Elliot Morley MP, Minister for Fisheries and the Countryside, Ministry of Agriculture, Fisheries and Food (MAFF). These witnesses covered the whole spectrum from science through producers and processors to the supermarkets, together with those responsible for certification, regulation and finally Government policy. We wish to thank all who gave evidence, either orally or in writing. We also ventured beyond Westminster, despite the best efforts of the railways to stop us, to visit two high profile and successful organic farms, one run by Helen Browning at Bishopstone in Wiltshire and the other by the Prince of Wales at Highgrove in Gloucestershire. In October we spent a day in Leicestershire, visiting the CWS farm at Stoughton, where organic, conventional and "integrated" farming methods are operated side-by-side, and two much smaller farms at Long Whatton and Normanton on Soar. We are extremely grateful to those whom we met during these visits for their frankness and for the many lessons we learnt about the practicalities of organic farming. This is the end.

9. Our specialist advisers for this inquiry were Professor Sir Colin Spedding, former chairman of UKROFS, who generously reprised his role as adviser to the Lords Committee on organic farming, and Professor Michael Haines of the University of Wales, Aberystwyth. We hereby express our appreciation of their assistance and guidance throughout the inquiry.

Structure of Report

10. The Report is structured as follows. Section II examines the expansion of organic farming and of the demand for organically produced food in the UK. Section III looks more closely at the supply chain and the role of supermarkets and producer-owned co-operatives. Section IV looks at the system for certifying organic products and for setting standards and considers possible improvements to that process. Section V and Section VI cover the role of Government in supporting organic farming, first through direct financial assistance and second through other, indirect means such as research and development or training. Finally, Section VII sets out our main conclusions and recommendations.

II. EXPANSION OF ORGANIC FARMING

Organic production in the UK

11. Although there is some discrepancy in the figures supplied by MAFF and those supplied by the Soil Association, it is clear that there has been a major increase in the amount of land and number of farmers engaged in organic production in recent years. It is equally clear that the rate of growth is accelerating. At the end of 1996 some 50,000 hectares of land in the UK was being farmed organically (either fully organic or in conversion).[3] By April 1999 the figure was 276,000 hectares and by the end of that year it had reached 425,000 ha, or just over 2 per cent of all agricultural land in the UK.[4] The Soil Association, whose figures tend to be lower than those of MAFF, estimated that in April 2000 organic farming covered about 2.3 per cent of the total UK agricultural area, of which 0.6 per cent was fully converted.[5] The geographical distribution of this land is weighted towards the South and West and Scotland, reflecting the increasingly high proportion of organically farmed grassland (87 per cent of all organic land in April 2000 and 79 per cent in April 1999) compared with arable and horticultural production (13 per cent in April 2000 and 21 per cent a year earlier).[6]

12. Statistics on the number of registered organic producers also reveal significant increases: as of April 2000, there were 2,865 farmers licensed as in conversion or fully organic, compared with 1,568 in the previous year, 1,064 in April 1998 and 828 in April 1997.[7] Placed alongside the data for organic acreage, these figures illustrate a trend, identified by the Soil Association, towards the conversion of larger holdings. While there have been some very large-scale conversions of Scottish moorland and it remains the case that organic holdings in Scotland are significantly larger on average than those in the rest of the UK (330 ha compared with 116 ha in 1999), the average farm size is on the increase, even discounting the Scottish figures.[8] The conversion period before produce can be marketed as organic varies between sectors (24 months for crops and grassland; 36 months for perennial crops)[9] but it inevitably involves a timelag in increases in the acreage of organically farmed land feeding through to the production of organic foods, so the full effect of these changes is yet to be felt. Nevertheless, organic primary production in the UK grew by 25 per cent in the year from April 1998 to reach a total of £50 million (0.4 per cent of the total UK agricultural output).[10] We turn now to the individual sectors within that global picture.

Meat

13. In 1998/99, the latest year for which statistics are available, the farm gate value of UK organic meat reached £7 million, an increase of £2 million on the previous year's figures.[11] This total included 4,200 cattle, with beef production increasing by 24 per cent on 1997/98, 26,000 sheep (representing an increase in lamb production of 13 per cent), nearly 200,000 table birds (up 52 per cent) and 12,000 pigs (an increase of 71 per cent).[12] Table 1 below sets out these increases in more detail.

---

[3]Ev. p. 131, para 7.
[4]*Ibid.*
[5]Ev. p. 99, section 3.1.
[6]*Ibid*; The Organic Food and Farming Report 1999 [SA (1999)], p. 8.
[7]Ev. p. 100, section 3.1; SA (1999), p. 7.
[8]SA (1999), p. 7.
[9]UKROFS Standards, January 1999 edition, 2.3.
[10]Ev. p. 131, para 7.
[11]Ev. p. 132, para 8.
[12]Ev. p. 234, para 5; Ev. p. 100, section 3.1.

Table 1: Organic meat production

Product	1997/98		1998/99	
	Number of animals/birds	Farm gate value (£m)	Number of animals/birds	Farm gate value (£m)
Beef	3,400	1.96	4,200	2.5
Lamb	23,000	1.04	26,000	1.4
Pork	7,000	0.96	12,000	1.8
Poultry (table birds)	125,000	0.77	190,000	1.2
Total		4.73		6.9

Source: SA (1999), p. 8.

It is expected that future increases in the livestock sector will dwarf those already seen, with the Meat and Livestock Commission (MLC) estimating that the number of cattle will increase by 300 per cent and of sheep and pigs by 500 per cent by 2001.[13] Eastbrook Farm Organic Meats Ltd, one of the largest players in the UK organic pig market, anticipated a tenfold increase in business by the end of 2004.[14]

14. Eastbrook Farm also expected that in four years' time it would be exporting "in reasonable volumes".[15] In most livestock sectors, however, there will be difficulty in meeting domestic demand. The MLC stressed that even if its predictions for 2001 were borne out, organic livestock would still account for under one per cent of total production in Great Britain and would be unlikely to rise above five per cent by 2010, unless cost differentials were reduced.[16] Although MAFF told us that the UK is currently "self-sufficient in most organic meats",[17] the MLC indicated that imports were increasing, with beef coming in from South America and beef and pigmeat from Scandinavia and Northern Europe.[18] ASDA singled out home-reared organic meat as the clearest illustration of the shortfall in British product.[19] At the moment, ASDA is able to source only seven beef cattle per week from the UK whereas the supermarket predicts a future demand for 140 head of organic beef a week, that is over 3,000 more than the current UK annual total output.[20] Overall, in 1999 UK producers took a 95 per cent share of the retail market for organic meat[21] but this position looks decidedly shaky. Constraints on expansion specific to this sector include the conversion time (a minimum of four years for organic beef production); the cost of rearing organic dairy calves; and the costs of conversion and ongoing costs for pig production.[22] The impact of the scarcity of organic animal feed is also a factor (see paragraph 17 below).

Dairy

15. Organic milk production in 1998/99 was 17 per cent higher than the previous year, with output of 27 million litres and a farm gate value of nearly £8 million.[23] The UK is largely self-sufficient in organic liquid milk but the demand for processed dairy products means that around 40 per cent of the total organic dairy sector is met by imports.[24] Of all the agricultural sectors, the organic milk producers have perhaps been the most willing to work together through the

---

[13]Ev. pp. 234-5, para 5.
[14]Ev. p. 152.
[15]Ev. p. 153.
[16]Ev. p. 235, paras 5-6.
[17]Ev. p. 132, para 8.
[18]Ev. p. 236, para 31.
[19]Ev. p. 223.
[20]Ev. p. 233.
[21]SA (1999), p. 21.
[22]Ev. p. 235, paras 7-8.
[23]Ev. p. 132, para 8; SA (1999), p. 8.
[24]Ev. p. 132, para 8.

establishment of the Organic Milk Suppliers Co-operative (OMSCo), which has negotiated rolling contracts with supermarkets such as Sainsbury's or with processors such as Yeo Valley Organic Company Ltd. The success of the latter company can be demonstrated by the fact that organic yogurt now accounts for 6.3 per cent of the total UK yogurt market[25] and illustrates the need for dedicated processing capacity.

Eggs

16. The organic egg industry has outstripped the rate of expansion experienced by other agricultural sectors with a doubling of production between 1997/98 and 1998/99 in terms of number of eggs and farm gate value, now 75 million and almost £6 million respectively.[26] The UK meets 100 per cent of domestic demand for organic eggs, although as yet this represents only 2 per cent of the total egg market.[27] As with the meat production sector, the organic egg industry faces a new challenge in the form of the EU organic livestock regulations which came into force on 20 August 2000. The Soil Association predicted that this might "result in a short-term reduction in organic poultry enterprises as producers adapt to meet the new standards".[28] We discuss the issues raised by the livestock regulations in more detail below (see paragraphs 57 to 60).

Arable crops

17. Organic arable crops – mainly wheat (58 per cent of the organic cereal sector), oats (20 per cent) and barley (10 per cent) – reached an estimated farm gate value of £6 million in 1998/99, with production just 19 per cent higher than in the previous year. This relatively small increase reflects the lack of interest shown by arable farmers in organic conversion and perhaps the greater difficulty in achieving conversion, compared with farmers in other sectors. Shortfalls in arable crops have an unfortunate knock-on effect on the availability of organic animal feed for the livestock sector, making witnesses identify it as one of the key problem areas.[29] Only 30 per cent of UK demand for organic cereals is met by domestic supplies.[30]

Horticulture

18. The picture is still worse for horticulture which accounted for just 5 per cent of organic land in 1998.[31] Organic vegetable production saw year-on-year growth of 13 per cent to a total value of £18 million in 1998/99.[32] Most of this was potatoes, followed by other root crops.[33] The fruit sector lags even further behind, showing an increase of 5 per cent in farm gate values from 1997/98 to 1998/98 to reach £2 million. Both in terms of value and tonnage, the sector is dominated by apples, although even here we were told that 90 per cent of organic apples sold in the UK were imported.[34] Particular problems facing horticulture specialists wishing to convert include both the cost and technical difficulty of doing so.

Processed products

19. As with conventional foodstuffs, most organic food is processed or packaged in some way before it reaches the consumer. This part of the supply chain has seen similar growth to primary production in recent years, with an increase from 500 to 800 in the number of businesses licensed for the processing of organic foods in the UK from the end of 1997 to April 1999.[35] These 800 businesses accounted for nearly 2,500 individual enterprises engaged in a variety of businesses, as shown in table 2 below.

---

[25]Ev. p. 61.
[26]Ev. p. 132, para 8; SA (1999), p. 8.
[27]Ev. p. 238.
[28]SA (1999), p. 8.
[29]Q 184; Q 444.
[30]Ev. p. 132, para 8.
[31]*Ibid.*
[32]*Ibid.*
[33]SA (1999), p. 11.
[34]SA (1999), p. 11; Ev. p. 185, para 13.
[35]SA (1999), p. 17.

Table 2: Licensed organic processors by enterprise type and location, April 1999

Type	North	Mids	Scotland	Wales	NI	South	West	East	Total
Abbatoirs	7	8	2	7	3	4	14	6	51
Animal feed	0	11	0	2	0	6	12	3	34
Baby food	0	3	3	0	0	3	3	2	14
Beverages	4	8	2	2	0	19	14	12	61
Breads/baked	11	26	4	3	0	27	11	2	84
Cereals/cereal products	14	27	10	8	0	20	12	27	118
Chilled/frozen	3	8	0	2	0	8	3	4	28
Conserves,etc	14	15	4	10	0	20	15	10	88
Distribution and packing	71	98	28	66	3	124	102	111	603
Dairy processing	3	22	15	4	0	28	43	2	117
Dried goods	2	8	3	2	0	11	3	8	37
Exporting	0	2	0	2	0	4	0	2	10
Fish	0	2	3	3	0	0	0	0	8
Fresh foods	47	155	46	71	6	153	146	139	763
Importing	22	30	12	2	0	88	27	44	225
Meat processing	10	16	4	10	2	8	26	14	90
General food processing	22	26	8	11	0	22	18	26	133
Total	230	465	144	205	14	545	449	412	2464

Source: SA (1999), p. 18.

For the year to April 1999, the UK organic processing market was worth over £300 million (domestic plus imports), of which £135 million was fruit and vegetables.[36] Intensive processing of organic foods is a matter of some argument within the organic movement, as the philosophy advocates simply prepared foods whilst consumers often seek organic equivalents of conventional processed foodstuffs.[37] However, there is widespread willingness among processors to supply the growing demand for organic products.

Retail market

20. The expansion in organic production is racing to keep up with the growth in customer demand. In April 2000 the total organic retail market was worth almost £550 million on an annual basis and no-one has seriously questioned predictions that it will reach £1 billion next year.[38] Figure 1 below shows the trends in market growth since 1993.

---

[36]SA (1999), p. 18.
[37]Qq 521-522.
[38]Ev. p. 100, section 3.2.

Figure 1: UK retail market growth – actual and projected

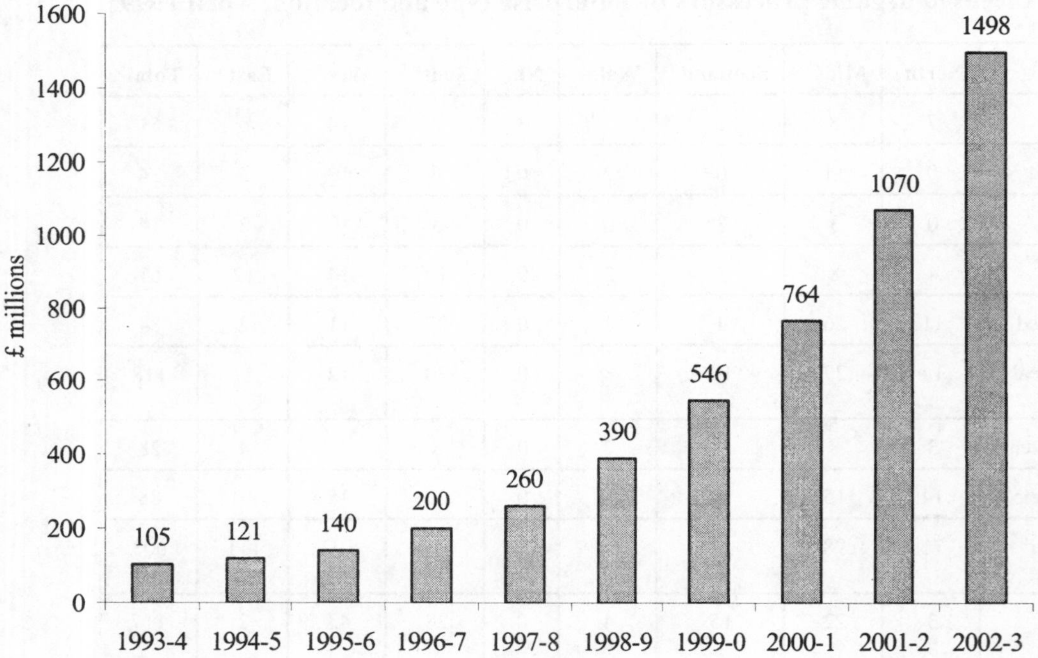

Source: SA (1999), p. 22.

To keep these figures in perspective, in 2000 organic food accounted for just 2.5 per cent of the total UK food market. To reach the market share of 20 per cent sought by the Soil Association by 2005, would require annual growth of 50 per cent which, of course, is much harder to achieve when the amounts involved are already significant. Doubts have been expressed about the ability of the market to sustain such growth; for example, the MLC cited a survey of multiple retailers which revealed expectations that the market for organic meat would command between 3 and 10 per cent of total meat consumption in five years' time.[39] Nevertheless, the supermarkets are gearing up for further strong growth over the next few years: Sainsbury's predicted that the market will "begin to level out" in 2003 or 2004 and would "reach a peak at the 10 year mark by 2010".[40]

21. The major supermarkets have responded to consumer demand by stocking vastly increased ranges of organic foods. Sainsbury's now offers over 630 products, including frozen meals, vodka and pet foods.[41] By comparison, in 1996 it offered just 42 lines, covering basic commodities and fruit and vegetables.[42] Despite the cornucopia of products now on offer, it is still the fruit and vegetables sector which dominates the organic market, as figure 2 below illustrates.

---

[39]Ev. p. 235, para 10.
[40]Ev. p. 73, para 3.5.
[41]Ev. p. 73, para 3.3.
[42]Ev. p. 73, para 3.2.

Figure 2: Retail value of UK organic market, 1998/99

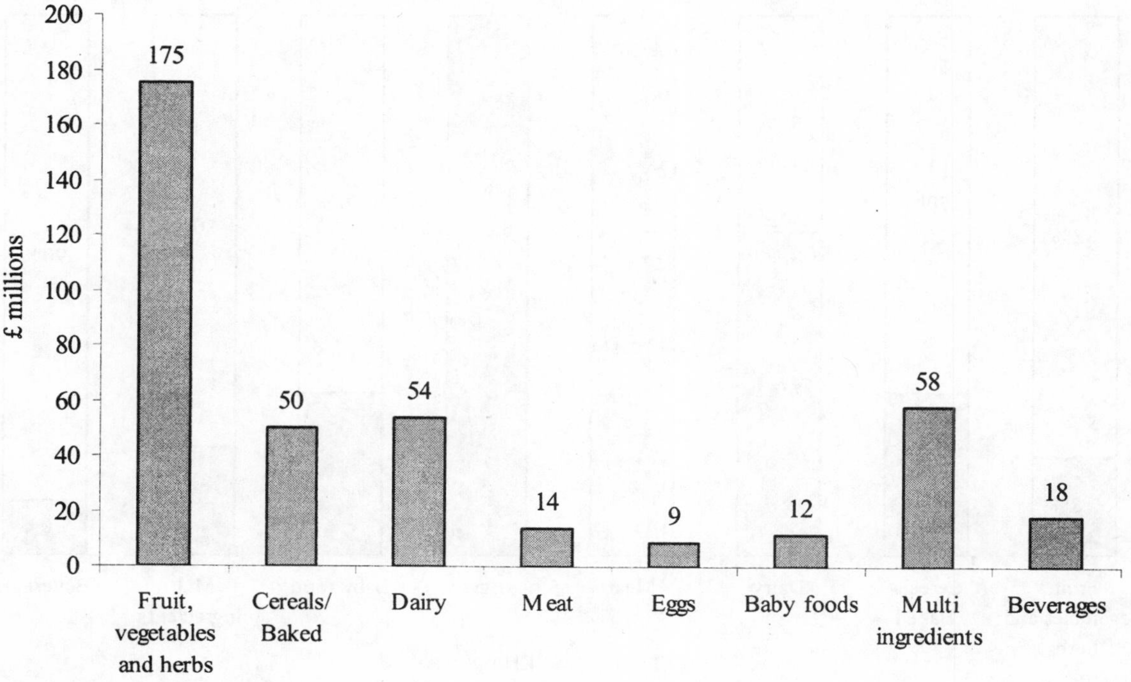

Source: SA (1999), p. 22.

The biggest expansion in the future is likely to come in baby food where organic products are variously predicted to take 40 per cent of the market by the end of 2000 and 100 per cent by the end of 2001.[43]

22. The "organic trade gap" is set to continue to grow as demand outstrips the rate of conversion. It currently stands at around 70 per cent, although this varies between sectors (see figure 3 below).

---

[43]Ev. p. 73, para 3.1; Ev. p. 100, section 3.2.

Figure 3: Imports of organic food (% share), April 1999)

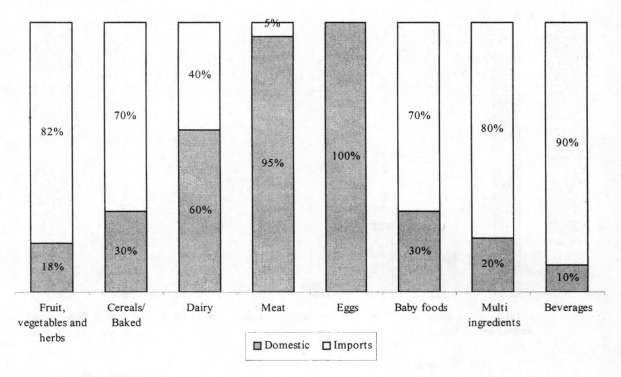

Source: SA (1999), p. 21.

Given the dominance of fruit and vegetables in terms of overall sales, the large percentage of imports in that sector plays a major part in setting the average for the whole organic market. This indicates the need for some caution in assessing the trade gap in organics since many of these products are inevitably sourced from abroad, for example, oranges, mangoes or bananas. By comparison, in 1999 home production for the conventional market accounted for 71 per cent of vegetable supplies but only 11.9 per cent of fruit. Imports are not by definition a bad thing: the NFU argued that they were crucial to developing the market at this stage before UK farmers had had time to complete conversion.[44] **Nevertheless, it is clear that there is a huge opportunity for UK producers to expand still further into organic farming to meet a ready market.**

International comparisons

23. Brookes West economic consultants estimated that the current size of the global organic market at retail level is about $20 billion.[45] Table 3 below sets out figures for key world markets since 1997.

[44]Ev. p. 36.
[45]Ev. p. 177.

Table 3: Key world markets for organic food and drink (US$ m)

Country	1997	1999 (1)	2000 (2)
US	4,200	6,000	8,000
Japan	1,000	2,000	2,500
Germany	1,800	2,500	2,500
France	720	1,000	1,250
UK	450	650-700	900
Netherlands	350	400	700
Switzerland	350	N/A	600
Denmark	300	N/A	600
Sweden	110	140	400
Italy	750	N/A	1,100
Austria	225	300	400
Other EU	200	N/A	500

Notes
1. Estimate
2. Forecast
3. N/A = not available
Source: ITC, USDA, Economist (Ev. p. 178).

It can be seen from this table that the value of the UK organic market is lower in absolute terms than in several other EU Member States. Denmark, Sweden, Germany and Austria are examples of countries where organics enjoyed a larger market share of domestic consumption than in the UK.[46] These countries all have a higher relative proportion of organic farms and therefore higher production levels than the UK (see table 4 below). Sweden, missing from the table, had 11.2 per cent of its agricultural land in organic production in 1999/2000.[47]

Table 4: Organic food and farming in selected countries in Europe, 1998/99

Country	Population (m)	Organic retail sales	Organic and in-conversion land area (ha)	Number of organic farms
Holland	15.5	£130m	20,000	800
Denmark	5.3	£220m	98,000	2,100
Germany	82	£1.5bn	430,000	9,300
Austria	8	£180m	345,000	19,200
France	59	£400m	220,000	6,300
Italy	57	£15m	830,000	30,000
UK	59	£390m	240,000	1,568

Source: SA (1999), p. 3.

[46]Ev. p. 18, para 6.
[47]Ev. pp. 175-6, table 1.

Factors affecting expansion

Reasons for expansion

24. Organic production in the UK is chasing the booming consumer demand for organic food. The reasons for this demand obviously vary between individuals but some clear themes emerge. One is food quality; for example, the perception that properly-cured organic bacon tastes like "real" bacon. A second, and probably the strongest, theme is concern about food safety, driven by food scares such as *E. coli*, Salmonella, BSE and genetically modified foods, as well as by growing awareness of the pesticides and other chemicals, including antibiotics, regularly used in the production of conventional foodstuffs. These scares have increased the desire of consumers for knowledge about the food they eat. Similarly, growing interest in the environment, in animal welfare and in other ethical issues linked to food and farming have heightened the attraction of organic foods because of the values which they appear to represent. In addition, there are nutritional issues as some consumers perceive organic foods to offer health benefits over conventional products.

25. Another reason for the expansion of the retail market may be an economic one. Brookes West pointed out that there is "a strong link" between organic consumption and high GDP levels.[48] They also raised the interesting example of Austria where the rapid development of the organic market coincided with accession to the EU: at that time prices of foodstuffs in general fell with the removal of agricultural protection, with the result that "organic produce was sold at prices that many consumers were used to paying in the past for conventional food".[49] It is possible that similar factors are at work in the UK where conventional food in many sectors is now selling for less in real and absolute terms than it was a few years ago, thus allowing consumer to choose "to forego a cheaper alternative ... rather than ... actively choosing to pay more" in buying organics.[50] The supermarkets were keen to assure us that the whole range of their customers is now choosing organics and that the stereotype of the organic customer as middle class, urban and from the south east is no longer relevant. This change is as likely to result from the lower prices and higher visibility of organics as from other considerations.[51]

26. For farmers, the motivation behind organic conversion is primarily the attraction of commanding a premium price in a buoyant market, especially when set against the income problems in conventional agriculture. This is helped by the availability of Government grants. Farmers also share many of the consumer concerns referred to above as well as an interest in a more "natural" way of farming. Some see organic farming as a means of adopting greater extensification of production whereby they feel that they have more control over what they do and what they produce. Mr Watson, a farmer from Devon, explained that his family had converted "as most people do, for a combination of ideological and commercial reasons".[52] A processor with many years' experience of the industry added that if farmers "do get in it just for money, I think some of them will learn their lesson very quickly and come back out".[53] Organic farming was "a way of life",[54] as much as a route to a profitable business. While consumers may be buying into the whole dream of organic farming when they purchase organic foods in a supermarket, farmers are finding that conversion involves a change of attitude not just of practice. The philosophy of organic farming is clearly a key factor in conversion.

Prices and premia

27. The high quality image of organic food is generally supported by high prices, offering producers a significant premium over conventional equivalents. The level of premium varies from sector to sector. For cereals and vegetables, it is around 50 to 60 per cent; for meat and

---

[48]Ev. p. 179.
[49]*Ibid.*
[50]*Ibid.*
[51]Ev. P. 170, para 3.4.
[52]Q 249.
[53]Q 344.
[54]*Ibid.*

dairy products, around 15 to 20 per cent.[55] The NFU attributed this difference to the balance between supply and demand for the product;[56] and it is certainly arguable that the premia have often been above the level needed to recover the additional costs involved in organic production and marketing.[57] We understand that premia are much lower (10 to 30 per cent) in more mature organic markets overseas[58] and recently we have also seen significant decreases here in supermarket prices. For example, ASDA has "an aspiration of [prices for organics] being no more than 30 per cent higher than the non-organic equivalents",[59] while Iceland has gone one better by announcing the wholesale conversion of its own-brand vegetables to organic without charging a premium. In order to do so, it has bought up 40 per cent of world supplies of organic vegetables, a further move away from the notion of organic farming as localised production.

28. These moves by the supermarkets have caused some concern among organic farmers who are anxious that the power of the multiple retailers could push producer prices down below sustainable levels. The fear is that the *actual price* received for organic produce will fall rapidly in line with the increase in supply and that the supermarkets will fail to keep their promise of paying for consumer reductions through economies of scale or out of their own profits rather than those of their suppliers. The difference between price and premium is a vital one. Organic farmers receive a price which may represent a premium over conventional products, but what they get is a price, not a guaranteed premium. For example, the ex-farm price of organic milk (c. 29.5 ppl) is significantly above the current ex-farm price for conventional milk (c. 18ppl) but not so much higher, particularly when the additional costs of production are taken into account, than that obtained by conventional producers in 1995 (c. 24.9 ppl). Mr Tucker of Yeo Valley Organic Company told us that the company had "been trying to get that message through to the farmers that it is not a premium, if the conventional price does start to rise again it will not push up the organic price".[60] The farmers we met recognised this fact and were seeking stability in contracts and pricing, rather than a guaranteed premium over conventional producers.

29. It is likely that over time, even without pressure from the supermarkets, organic prices will fall. Mr Finney of Eastbrook Farm Organic Meats reminded us that at the moment "we have virtually no effective economies of scale to play with in a business of this size, in terms of transport, feed, packaging".[61] As the market grows and supplies grow, such economies of scale will be possible throughout the industry, a point made strongly by Iceland.[62] This might not affect farmers directly: Dr Lampkin was not convinced that "there will be a significant reduction in prices to farmers in the short to medium term", ie. three years, because increased production was likely to replace imports rather than add to the total supply, the market was continuing to expand and "there is a lot of potential for significant economies of scale in the processing and distribution system".[63] He also believed that farmers could afford to see their profits for organic produce cut,[64] although we tend to agree with the Minister, Mr Morley, that it is justified for those who wish to support organic farming to pay a premium to cover its extra costs.[65] We believe that as the market develops there will be some erosion of the current advantage over conventional prices paid to producers (not least because conventional prices may themselves recover as we have seen, albeit hesitantly, in milk) and that any farmer going into the business now would be wise to plan for prices lower than those that organics have commanded in the past.

Constraints on expansion

30. We have already referred to some of the constraints on expansion of UK organic production within individual sectors, such as conversion times or technical difficulties in organic

---

[55]Ev. p. 31.
[56]Ev. p. 31.
[57]Ev. p. 179.
[58]Ev. p. 185, para 16.
[59]Ev. p. 233.
[60]Q 344.
[61]Ev. p. 152.
[62]Ev. p. 78, para 5.4.
[63]Q 125.
[64]*Ibid.*
[65]Q 597.

practices. There are, however, some constraints which are common to the whole organic farming industry. An obvious example is that Government funding for organic conversion is limited both in the overall budget and in the need for land to be IACS-registered before it can receive OFS assistance, an additional layer of bureaucracy for those sectors which would not otherwise be concerned with IACS. In addition, for nearly a year, Government funding has not been available at all (although we note that early organic producers, like Mr Watson, did not require this incentive). Other problems could be defined as gaps in the infrastructure, such as shortages in animal feed or organic seed; the declining number of small abattoirs (essential to the organic livestock industry); the shortage of processing capacity in general; or insufficient training.[66] There is also a need for more technical research to help farmers and growers address the consumer demand for a wide variety of choice in each food type, together with cosmetically attractive produce and year-round availability. This is again closely linked to the requirement for better marketing in order to meet or educate customers' expectations.[67] Finally, there is the impact of regional clusters on the development of organic farming. Where these emerge or are encouraged, they may create a dynamic towards even greater growth; where they are absent, expansion may be constrained in those geographical areas.

31. The constraints outlined above can be addressed by the industry and Government and we will discuss later in this Report strategies for doing so. However, there is another set of constraints which are less amenable to practical solutions. The organic boom is highly dependent upon consumer perceptions and is equally vulnerable to changes in those perceptions. There are two main threats here. The first is that the rapid growth of the market might encourage standards to be eroded, whether of imported produce where controls are inevitably more difficult to enforce or of domestic produce. If this happens and becomes a public issue, then consumer faith in the organic industry will be shaken and there could be a rapid fall in its fortunes. Even worse would be a major food scare involving organic foods. This underlines the importance of the regulatory approval and inspection systems in ensuring food safety, product authenticity and reassuring consumers that they are getting what they are paying for.[68]

32. The second threat, however, is that consumers will cease to be prepared to pay for organic foods. The impact of food scares may fall away; the general public might come to doubt the claims made for organic in the context of food safety or quality; or difficulties might arise over the popular perception of organic as chemical-free or environmentally benign. Public attitudes towards GM foods might also alter over time, with a potential impact upon the market for organic and conventional products.[69] **It is therefore vital that the organic industry develops its ability to market its products effectively so that they appeal not to sentiment but to proven benefits. The industry may need to be less messianic and more marketing-orientated in its public presentations.** The hard core which has always purchased organic food is very small; to appeal to the majority market beyond that in order to ensure a viable volume sector into the future, the organic movement must demonstrate the qualities which differentiate its products from conventional ones and must continue to respond to the market in offering the service and range demanded by consumers.

III. THE SUPPLY CHAIN

Direct marketing schemes

33. The consumer demand for organics has been fed in part by the growth in distribution systems dedicated to or geared towards the organic sector, such as farmers' markets and box schemes. These methods adhere to the traditional ethos of organic production with its emphasis on local production and unprocessed, unpackaged products. Together with farm shops and internet services, such farmer-run direct marketing schemes are seen to deliver benefits to the farmer in terms of increased profit margins, to the local community in terms of employment

---

[66]Ev. pp. 174-5, 4.1; Ev. p. 163, section 6.
[67]Ev. p. 32.
[68]Ev. p. 180.
[69]*Ibid.*

opportunities and to the consumer in terms of fresh organic food and immediate traceability.[70] In 1998/99 £58 million worth of organic food was sold at the farmgate, through box schemes or from farmers' markets,[71] and the figures are likely to be much higher this year. Farmers' markets are now established in many major towns and cities, bringing largely local produce within the reach of more consumers. Moreover, it is clear that there is room for this sector to expand as box schemes are often oversubscribed. The Government has been enthusiastic about these developments in marketing by producers. We too believe that they should be encouraged, perhaps through the schemes available under the England Rural Development Programme. **We recommend that the Government encourage the further development of local marketing schemes, such as farmers' markets and box schemes, through the provision of advice and ERDP funding.**

Supermarkets and organics

34. The greatest growth in organic sales has been through the main supermarkets. More customers will have access to these outlets on a national basis than can be reached by other distribution methods. The multiple retailers have a 69 per cent share of the organic retail market with sales of £269 million in 1998/99[72] and given their stated commitment to organic produce, their dominant position seems assured. The supermarkets have been accused of overstating the claims for organic food, particularly in terms of food safety and animal welfare, but they have put huge efforts into marketing organic produce and into meeting the demands of their customers for expanded ranges and organic versions of conventional products. However, these efforts have not removed all suspicions within the organic sector of the motives of the multiple retailers and of their ultimate impact upon the development of organic production in the UK. The basis of these concerns is both specific to the organic sector, in that the whole concept of multiple retailers sits uneasily with the "purist" organic ethos of local food for local people, and general in the perception of many in the farming industry – organic and conventional – that the power of the supermarkets is detrimental to the interests of producers.

35. The organic industry as it is currently constituted causes significant distribution problems when set against the demands of supermarket customers. For example, it is often small-scale, leading to increased costs in processing and transport; production is often erratic or has accentuated seasonality; and the products cannot be guaranteed to meet tight specifications or to be cosmetically perfect, leading to considerable waste.[73] Research into technical issues may help overcome some of these difficulties but it is hard for supermarkets to plan their supplies on this basis. It is not surprising, therefore, that Iceland, for example, was encouraging "large players to enter the market and achieve the necessary economies of scale".[74] At the moment, and for the foreseeable future, Iceland is achieving its objective of selling organic food at conventional prices by importing a large percentage of its supplies. However, it is also understandable if many organic farmers see this as a threat to their livelihood in terms of the prices offered for their produce and of competition from much larger concerns with lower overheads. There are also fears that the "unseemly scramble" by retailers for available supplies in the current situation will have a detrimental impact on standards as producers rush to meet the demand, putting in jeopardy the high values on which the appeal of the organic sector rests.[75]

36. The supermarkets have worked hard to counter these concerns in a variety of ways, including funding for research and development as well as direct assistance for conversion.[76] In evidence to us they were keen to emphasise their commitment to long term partnerships with suppliers. For example, Sainsbury's is working closely with the Organic Milk Suppliers Co-operative (OMSCo) with which it has a five-year deal guaranteeing prices and volumes.[77] ASDA has similar arrangements with farmers who have signed up to its organic meat conversion

---

[70]Qq 303-4; Ev. p. 162, section 5; SA (1999), p. 23.
[71]SA (1999), p. 22.
[72]*Ibid.*
[73]Ev. pp. 35-36; Ev. p. 85; Ev. p. 236.
[74]Ev. p. 78, para 5.4.
[75]Ev. p. 236, para 22-4; Ev. pp. 152-3.
[76]Ev. p. 35.
[77]Ev. p. 72, para 2.6.

scheme,[78] while Iceland assured us that "Our relationships with our organic producers are totally different to our producers and suppliers of other products, simply because of what we are trying to do together".[79] There are also efforts being made to respond to the organic ethos by sourcing locally, perhaps to sell at a higher price, and by addressing issues such as excess packaging.[80] Nevertheless, suspicions persist that the supermarkets will try to impose their conventional trading practices upon organic suppliers, a concern fed by remarks such as that of Tesco that "supermarkets' traditional methods of sourcing and selling products will promote organic food".[81]

37. At the moment, the supermarkets must fight for scarce supplies of organic produce to meet the demands of customers. It is, therefore, not surprising that they are responding to this pressure by setting up special arrangements with the organic sector and that there are fears that such willingness to adapt to the traditional patterns of organic production will not last once supplies are more in balance with demand. The organic sector itself can react to this situation either by turning away from the supermarkets, in which case it will lose much of its potential market, or by working with them to ensure that its needs and special qualities are recognised. The Soil Association took the view that "it is very important to build a dialogue with the supermarkets in parallel with their customers who are increasingly asking for more organic food".[82] To achieve this, the Association had established a multiple retailers' working group which provided a forum in which issues could be raised. We applaud this initiative and agree that "It is better to influence [the supermarkets'] practices in a better direction", than to be negative about the role of the multiples.[83] There will always be a niche market for organic producers who want to sell their output through local or direct marketing but the large scale expansion of organic production depends upon the supermarkets. Their need to respond to customer demand should ensure that they pay more than lip service to maintaining their contracts with organic suppliers and to the need to keep standards high and prices realistic. Moreover, we expect to see them developing the factors which make organic produce unique, such as variations in size and specifications, into an advantage as consumers continue to express a desire for more "wholesome" foodstuffs for which certain segments of the market are willing to pay a premium. **Supermarkets will be the main, although not the only, distribution channel for organic produce. It is critical that they are involved in the design of and encouraged to co-fund future initiatives to further organic conversion.**

Supplier partnerships

38. An obvious answer to the difficulty of the mismatch between the need for economies of scale in dealing with supermarkets and the small-scale nature of much of the organic production sector is the development of supplier partnerships or co-operatives. These arrangements strengthen the hand of the farmer, who is no longer working in isolation and can make use of common facilities and combined negotiating power, and are also welcomed by the supermarkets who find it easier to contract with a group of farmers than with individuals. There have been notable successes in creating such co-operatives within the organic sector. OMSCo was originally set up by a groups of organic farmers in response to the lack of processing capacity for organic milk.[84] It is now the major player in the organic milk sector, acting as a link between the processor or retailer to the farmer both for contracts and for development. For example, Yeo Valley Organic Company is working with OMSCo to encourage farmers to convert.[85] Similarly, in the livestock sector, the Organic Livestock Marketing Co-operative was set up some five years ago by the Soil Association, Organic Farmers and Growers Ltd and Eastbrook Farm Organic Meats "to create an orderly marketing environment for primary producers, with ex-farm prices fixed for long periods".[86] It may be harder for other sectors to develop such partnerships but we agree with the NFU that "there are many opportunities for co-operatives and other farmer-

---

[78]Ev. p. 224.
[79]Q 454.
[80]Qq 435, 410, 434.
[81]Ev. p. 213.
[82]Q 540.
[83]Q 540.
[84]Q 312.
[85]Q 343.
[86]Ev. p. 235, para 19.

controlled businesses to flourish in the marketplace".[87] The Minister for Agriculture, Fisheries and Food has often expressed the view that such co-operatives should be encouraged in agriculture more generally. We believe that there is an opportunity here for the Government to assist the development of farmers' co-operatives in the organic sector through the ERDP and other funds aimed at rural development. **We recommend that the Government work with the bodies responsible for the promotion of organic production to ensure that rural development funds are channelled into the development of supplier partnerships and farmer-controlled co-operatives in the organic sector.**

Abattoirs

39. A particular obstacle to the expansion of organic livestock farming has been the availability of local abattoirs. Processing capacity for organic livestock is limited by organic protocols which require that animals are slaughtered at an abattoir registered and checked by a certification body. The closure of many small abattoirs over the last few years has left organic farmers in the position that the nearest available facilities may be many miles away. This is unacceptable in terms of cost (organic livestock farmers tend to operate on a small scale) and in terms of animal welfare.[88] Moreover, it could have a knock-on effect on consumer choice as the lack of locally-slaughtered organic meat could lead to the loss of independent butchers, farm shops, farmers' markets and other speciality outlets.[89] We know of at least one case where a farmer was forced to forgo his organic status because of the loss of a local organic abattoir and the importance of such facilities was stressed to us in informal discussions with others.[90] The NFU described the "lack of a local slaughtering facility" as "the top of the list" of problems faced by organic farmers.[91]

40. The recent spate of closures or threatened closures of small, local abattoirs has been attributed to the charging regime for Meat Hygiene Service inspectors.[92] The Government has tried to address this problem in the Rural White Paper, published in late November 2000, by promising to "target help for small and medium-sized abattoirs".[93] It announced "new, additional aid (worth £8.7m in 2001-02) in respect of meat inspection costs to help secure the future of small and medium-sized abattoirs".[94] **We welcome this additional aid and await with interest details of the package and we urge the Government to stimulate the development of new small abattoirs, including mobile abattoirs.**

IV. CERTIFICATION AND STANDARDS

The regulatory system

41. The harmonisation of the regulatory system for organic production throughout the European Union began ten years ago with the introduction of Council Regulation 2092/91 on *organic production of agricultural products and indications thereto on agricultural products and foodstuffs*. Coming into force in 1992, the Regulation set out rules for the production, processing, labelling and marketing of organic crop products.[95] The latest amendment to the Regulation (Council Regulation 1804/1999) brings livestock within its scope with effect from August 2000, thus closing a major gap in the EU's organic legislative framework. The standards prescribed by the Regulation are the minimum which must be met; where the Regulation is silent, the "competent authority" in an individual Member State (the establishment of which is provided for by the Regulation) may set its own standards. In addition, the competent authority is required to approve private sector certification bodies, which in turn may impose on their members standards higher than those in the Regulation. Products produced in accordance with

[87]Ev. p. 36.
[88]Ev. p. 155, para 11.
[89]Ev. p. 163, para 6.2.
[90]Ev. p. 250, annex; private information.
[91]Q 226.
[92]Ev. p. 155, para 11.
[93]Rural White Paper, *Our Countryside:the Future*, Cm 4909, p. 92.
[94]*Ibid.*
[95]Ev. p. 132, para 11.

the standards and inspection processes set out in the Regulation in any Member State may be imported into any other. In addition, imports from countries whose standards and inspection systems have been recognised as equivalent to those applying within the EU (Argentina, Australia, Hungary, Israel, Switzerland and the Czech Republic) are treated as if they were produced within a Member State, whilst produce from other third countries must be authorised by the competent authority in the importing Member State and the importer must be registered with an organic inspection body.[96]

42. In the UK, the competent authority is the United Kingdom Register of Organic Food Standards (UKROFS). UKROFS predates Regulation 2092/91, having been established in 1987, and has seen its role change over time, first with the introduction of the main Regulation and more recently with the agreement of the livestock regulations, which has removed from it the necessity to set its own standards in this area. The focus of UKROFS is now "on ensuring that organic certifying bodies correctly interpret and implement [the Community] legislation rather than actually setting standards".[97] UKROFS currently recognises eight inspection bodies as running certification systems in compliance with the Regulation: Organic Farmers and Growers Ltd (known by the designation on labels of UK 2), Scottish Organic Producers Association (UK 3), Organic Food Federation (UK 4), Soil Association Certification Ltd (UK 5), Bio-Dynamic Agricultural Association (UK 6), Irish Organic Farmers and Growers Association (UK 7), Food Certification (Scotland) Ltd (for organic salmon; no UK number allocated) and Organic Trust Limited (UK 9). The designation of UK 1 is reserved for operators who are directly certified by UKROFS itself because they wish to work only to the basic EU standards (accounting for just 15 enterprises at the present time).[98] The private inspection or sector bodies are responsible for ensuring that those registering with them are acting in accordance with the prescribed standards of that body, whilst UKROFS is responsible for ensuring that the private sector bodies are competent and fulfilling their certification and inspection procedures to the required level.

43. A number of issues were raised with us during the course of this inquiry as to the nature of the regulatory system for standards and certification, both as applied in the UK and on a more general EU-wide level. On certification, comments focussed mainly on the number of certification bodies and the comparison between organic certification schemes and farm assurance. In this section of the Report, we therefore deal with these issues first. However, we recognise that both concerns are closely linked to questions about standards, particularly the difference between standards adopted by the various sector bodies in the UK and in other Member States and elsewhere, so we discuss this question, followed by an examination of the advantages and disadvantages of a single international standard. Finally, we look briefly at the role of UKROFS.

The number of certification bodies

44. As we have seen, including UKROFS itself, there are currently nine certification bodies in the UK, each entitled to use its own logo on organic produce. The certification bodies are responsible for inspecting those producers, processors and importers licensed by them to ensure that correct standards and procedures are being followed. There can be no doubt that the certification bodies have so far done an impressive job in regulating the industry and in promoting both organic products and the organic philosophy. The Soil Association pointed out that the certification bodies "provide a highly cost efficient mechanism for control of this sector which would otherwise have to be undertaken by MAFF or its designated agents at much greater cost."[99] MAFF recognises this role in the form of an annual grant which is currently under review.[100] The number of such bodies results from the history of the movement, with the dominant player, the Soil Association, operating to more stringent standards than its counterparts who offer their members the basic UKROFS standards, and the Bio-Dynamic Agricultural Association at the pure end of the organic spectrum, relying as it does upon very natural methods

[96]Ev. pp. 135-6, paras 33-5.
[97]Ev. p. 122, para 3.
[98]Ev. p. 124, annex B; Ev. p. 123, para 7.
[99]Ev. p. 101, section 3.4.
[100]*Ibid.*

of farming, with small scale production and an affinity with the Steiner philosophical approach.[101]

45. The multiplicity of certification bodies has not caused a problem in the past but we found general agreement that it could become a hindrance to the development of the organic sector. UKROFS was the only dissenter from the view that the different organisations – "the plethora of organic accreditation agencies", as one witness described it[102] – could be confusing for customers, who could be led to question just how organic was organic, and indeed for farmers trying to choose between them.[103] This situation has worsened with the growth of the organic market, both in its extension to include consumers who have no in-depth knowledge of the issues involved, and in the pressures it places on the certifiers. We heard of at least two cases where importers or processors had become frustrated by the delay in gaining approval from one certification body and had turned to another instead.[104] This causes difficulties throughout the chain right up to the customer who may be used to purchasing products with a particular logo, and we were warned that the situation was likely to get worse as "more bodies are trying to muscle in or come in to take advantage of the business opportunities".[105] In short, the structures applying to a "cottage industry" might not be so acceptable in a mass market.

46. In some countries, a solution to this problem has been found in authorising a single certification body; for example, in Denmark and Holland.[106] In Ireland, all organic enterprises are certified by the competent authority directly. By contrast, in Germany the sector is even more fragmented. It is clear that it is much more efficient to have larger organisations which can absorb the sudden increase in demand for certification (and operate a single database, rather than eight separate ones[107]) and that it would be much simpler for the farmer, food industry and the consumer, if there were fewer bodies. In the past, this would have been extremely difficult to achieve because of the different standards operated by the various bodies in the livestock sector, but the introduction of the EU regulation on livestock with a common minimum standard has brought the standards imposed by the bodies into greater alignment. The certification bodies to whom we spoke were themselves in favour of "fewer certifiers working more closely together"[108] or even "coalescing certification bodies into a neutral UK-Organic-Certification-Limited type body"[109], always with the proviso that within this arrangement there would be room for opting to operate to higher standards and to differentiate foodstuffs produced in this way. There was little support for bringing the certification and inspection of organic enterprises under direct UKROFS control and in any case, we see no need for such a step. We recognise that the EU regulation obliges UKROFS to register any new body which meets the requirements and that it would not be possible to prevent the formation of a new certification body. **The multiplicity of bodies with their different standards and symbols is a significant weakness and we believe that the certification bodies should be encouraged by the Government in their efforts at closer co-operation, which may lead ultimately to mergers.**

Organic certification and farm assurance schemes

47. Another area in which we believe that there is room for consolidation and simplification is the cross-over between organic certification and farm assurance schemes which are also proliferating in the UK. We recognise that organic standards and schemes "represent some of the first and best developed farm assurance schemes".[110] However, the expansion of other schemes in conventional farming, and in particular, the efforts made to rationalise these schemes to encourage consumer awareness and trust and to reduce the burden of both time and money on farmers, has led to questions over whether there is a need for the two systems to operate in such isolation. There are clear advantages for both sectors if inspectors were authorised to verify

---

[101]Q 502; Ev. p. 181.
[102]Ev. p. 186, para 21.
[103]UKROFS Q 557; eg. Ev. p. 180, Ev. p. 18, para 13; Q 396.
[104]Ev. p. 208, paras 10-11; Ev. p. 78.
[105]Q 502.
[106]Q 502.
[107]Q 502.
[108]Q 504.
[109]Q 502.
[110]Ev. p. 19, para 16

both organic and conventional schemes. For farmers, particularly those in conversion, it would remove the need for duplicate inspections; for the administrators of the scheme it could reduce costs; and for the organic sector as a whole it would meet the objections of those who point out that while farm assurance schemes have independent verifiers, the organic certification bodies both set the rules and carry out the inspections.[111]

48. We were encouraged to hear that moves are taking place to take matters forward in this way. The MLC wrote of discussions "on adding bolt-on modules to existing implementation protocols",[112] whilst the Soil Association was looking to a future ideal of "a one-stop shop for farmers and one symbol that the consumer is going to have to take notice of".[113] We recognise that much work would need to be done on the standards for any joint scheme: at the moment, the two sectors have different priorities and cover different issues, as well as having different standards where their coverage converges. It should, however, be possible to work towards the adoption of core values, perhaps even including the environmental objectives proposed by conservation groups.[114] This might be achieved by setting a quality standard along the lines of ISO 2000 which could be met by different but equivalent production methods. In the short term, it may be necessary to concentrate on closer co-operation and inspections, both on farm and beyond the farm gate. Given the goodwill expressed to us by the Soil Association, we believe there is potential to bring the two sides together to great benefit. **We recommend that MAFF facilitate discussions between the farm assurance schemes and the organic certification sector with a view to ensuring agreement on common core values and inspection protocols and with the goal of a single inspection process and shared symbols.**

Setting standards

49. The problem of confusion for consumers and others is greater where different certification bodies operate to different standards. As we have seen, the basic standards are those set out in the EU Regulation which apply across Europe with some flexibility to allow for local conditions. UKROFS told us that the regime prescribed in the Regulation "was based largely on practices which were in place at the time the standard was drawn up throughout Europe", making it "an embodiment of traditional practice more than anything else".[115] To this extent, there is some justification in the complaint that the standards have no proven scientific basis. UKROFS itself is able to add to the basic regulations in setting standards for the UK as a whole, although it is legally obliged to register anyone who conforms to the EC regulation, and beyond that, the certification bodies are free to establish their own regimes. Increasingly, it seems that the certification bodies are taking on more of the work of standard setting. The Soil Association has a committee system to develop standards and is adamant that it should retain the ability to set its own standards for organic production in order to allow them "to evolve, as the organic market develops, as advances in technology emerge and as our understanding increases".[116] The other certification bodies, with the exception of the Bio-Dynamic Association, tend to follow the UKROFS standards and act merely as certifiers and inspectors.

50. Two issues were repeatedly raised in evidence to us about standards for organic production. The first was that the standards themselves were non-scientific, arbitrary and illogical.[117] Various examples were given, including the grazing of non-organic stock in cider orchards (where the grazing limit of 120 days is divided between the number of organic orchards on a holding; ie. with two, it is 60 days and with three, 40), allowable inputs in horticulture (where the active ingredients are the same in chemical and organic treatments, but the latter are applied in less easily regulated amounts) and the withdrawal periods for antibiotics in livestock, which are double the safety margin already built into the UK regulatory system for conventional agriculture.[118] There was also a feeling that the standards were moving away from practical

---

[111]Ev. p. 171, para 5.5.
[112]Ev. p. 236, para 28.
[113]Q 503.
[114]Ev. p. 227, para 4.4.
[115]Q 564.
[116]Ev. p 101, section 3.3.
[117]Ev. p. 154.
[118]Ev. pp. 182-3; Ev. p. 185, para 9; Q 87.

experience towards ideology.[119] To some extent, the Soil Association did not contest the charge that its standards lacked scientific rigour. The Director explained that the development criteria included "evidence-based and non-evidence-based elements", the latter being "gut feeling, intuition, ethical considerations, and consumer attitudes".[120] This raises questions about the need to establish a scientific basis for organic agriculture to which we return later in this Report.

51. The second issue raised was the charge that the UK certification bodies were "gold-plating" the regulations, with the result that the standards required of UK producers were higher than those elsewhere in Europe or third countries.[121] For instance, conversion periods are longer and certain pesticides may be used elsewhere but not in the UK.[122] This has obvious implications for competitiveness in that producers will have higher costs and, it is claimed, may discourage companies from investing in the UK.[123] Closely linked to the perception that the UK was over-implementing the regulation is the belief that products from other countries may not be reaching the basic standards at all, because of inadequate supervision of imports.[124] Several witnesses were concerned that this could lead to loss of consumer confidence and a decline in the market for organic products.[125] It is difficult to address this concern when the different standards used in producing "organic" products make meaningful international comparisons hard to achieve.

52. It is true that within the EU standards do vary, as the regulation permits derogations and can be adapted to meet local conditions such as climate and soil type.[126] It is, however, not the case that the UK is generally applying a more rigorous regime than that in other countries. Dr Lampkin thought "the UK is in the middle position" since "in some respects other countries have higher standards and in some respects we have higher standards".[127] He named Denmark as a country with particularly high standards.[128] Dr Lampkin also dismissed suggestions that the standards were policed less well in other EU states than in the UK, arguing that "most countries now have pretty good systems in place to ensure the EU regulations are interpreted effectively and there is a fairly consistent approach across the European Union".[129] The Minister supported this view with the argument that countries such as Sweden, Austria, Holland and Denmark were "very proud of the way they apply" standards on organic production.[130] He also emphasised the role of the European Commission in checking that the regulations were correctly implemented.[131] There is a committee set up under article 14 of the regulation for this purpose. We believe that it would be of great benefit if more information were gathered and made publically available on what was happening in other Member States. UKROFS agreed that "it would be helpful if UKROFS more proactively sought information about the implementation of standards in other Member States" but insisted that it lacked the money to do so.[132] **We recommend that the Government ensure that the European Commission reports regularly on the implementation of the regulation and actively encourage the European Parliament to monitor this implementation. The Government should produce a "Non Paper" for distribution at the Agriculture Council to further this end. We further recommend that MAFF be pro-active in drafting EU regulations and ensuring their scientific validity before they are written into law. MAFF should also, either directly or through UKROFS or the FSA, seek to monitor the effect of regulations to ensure that other public policy objectives are not compromised.**

---

[119]Ev. p. 95.
[120]Q 519.
[121]eg Ev. p. 210, section 3; Ev. p. 96.
[122]Ev. p. 247; Ev. p. 169; Ev. pp. 187-8, paras 34-5.
[123]Ev. p. 210, section 3.
[124]eg Ev. p. 154.
[125]Ev. p. 236, para 31; Ev. p. 152.
[126]Ev. p. 51, section 5.
[127]Q 151.
[128]Q 151.
[129]Q 151.
[130]Q 697.
[131]Q 697.
[132]Q 555.

53. With regard to imports from third country suppliers, there is even less general knowledge and information available about the standards applied in production. Although we have heard evidence from companies who have gone to great lengths to guarantee the integrity of their supplies – for example, Sainsbury's organic pineapples and Yeo Valley Organic Company's arrangements with the Soil Association to check fruit from South America[133] – doubts and suspicions are still strong among UK producers.[134] The Soil Association agreed that the mechanisms for establishing equivalence of standards and certification procedures in third countries were "not very thorough".[135] This is clearly a matter of some concern, particularly as it is generally impossible to detect differences between products produced organically and those produced conventionally through tests carried out by trading standards officers. **Unless these discrepancies are removed, there is a real danger that confidence in organic food may be damaged.**

A single organic standard

54. The obvious solution to the difference in standards applied by the various bodies in the UK as well as in the rest of Europe and the world would be a single international standard to which all countries would be expected to adhere in order for products to be labelled organic. This has the advantage of simplicity, at least as a concept. It would end consumer confusion and reduce what the Provision Trade Federation described as "tensions between national producers and importers, each claiming that they are suffering a competitive disadvantage because they have to meet stricter standards."[136] There is also a standard ready to be adopted in the form of that developed by IFOAM (the International Federation of Organic Agricultural Movements). Fifteen major international certifiers including the Soil Association are currently accredited to the IFOAM Basic Standard[137] and three of our witnesses strongly supported its adoption.[138] The most cogent argument was that IFOAM met the criteria of a single standard: "it is a clear standard, it is understood by everybody and it is policed and audited to make sure it is followed through".[139] A further consideration was IFOAM's independence.[140] An alternative would be adopting the EC standards worldwide, perhaps redesignating them as obligatory standards rather than minimum ones.[141]

55. The battle between the proponents of high IFOAM standards and lower EU ones is not the only issue which would have to be decided in adopting a single standard within the UK, EU or the world marketplace. There is also the need for flexibility in that whatever standard were agreed, it would have to allow for adaptation to different geographical and climatic conditions if it were not to be seen as a unfair barrier to the development of the market. For example, requirements set down for livestock production in Denmark might not be suitable for farming in the Highlands of Scotland. A single standard would also have to allow for amendment and adaptation as more became known about organic farming or new products or techniques were developed. This might be difficult if it required agreement by many different interest groups. It must also be recognised that higher or alternative standards are a marketing tool. There will always be groups of producers willing to strive for higher and more demanding standards if this differentiates their products and guarantees them a market. We note too that with a single standard, the certification bodies would lose their responsibility for standard-setting, leaving them as mere checkers of compliance with the IFOAM or EU rules.[142] The Soil Association supported the IFOAM accreditation programme as one "designed to create a level playing field of standards and certification competence globally"[143] but believed that the loss of its own standards development function would "solve the problem of consumer confusion but it would

[133]Ev. p. 75, para 5.10; Q 389.
[134]Ev. p. 32.
[135]Q 527.
[136]Ev. p. 208, para 8.
[137]Ev. p. 73, para 4.3.
[138]Q 390; Q392; Q 442.
[139]Q 442.
[140]Q 443.
[141]Ev. p. 208, para 9.
[142]Q 390.
[143]Q 527.

also stifle the ability of the organic standards [bodies] to be involved."[144]

56. We agree with HRI that "There is an urgent need for a clear, defined understanding and pronouncement to producers, marketers and consumers of what constitutes 'organic produce', both to a realistic UK national standard, and for acceptance of organic food produced overseas and imported."[145] **We believe that IFOAM accreditation has much to offer in gaining acceptance for the standards met by imports from third countries and that the Government should support its widespread adoption.** However, on balance, we would not support the imposition of a single fixed standard upon the organic sector. The Soil Association told us of its plans to work with the other sector bodies in standards development,[146] which together with the new EC livestock regulations, should see a convergence of standards within the UK. Standards emerging from this new model would have to be endorsed by UKROFS, which would be responsible for ensuring that the standards were not so high as to undercut the competitiveness of the UK industry. The Government would also have a role in ensuring that the standard-setting procedure was properly resourced. The key to any standard – whether single or part of a whole range – is effective and uniform enforcement. Given the experience with the differential enforcement of EU rules, it is difficult to assume with confidence that a single international standard would, in practice, lead to a uniform quality of product. **We recommend that the Government endorse the involvement of the certification bodies in setting standards, with UKROFS acting as a check and balance in the system, and that the Government provide sufficient funding to ensure the rigour of standard-setting procedures.**

EU livestock standards

57. The new European rules for organic livestock production are a good example of the problems which may emerge when trying to set standards across the EU. The rules, adopted in August 1999 and coming into force in August 2000, allow temporary derogations in some areas as well as permitting Member States to apply stricter standards to their own producers. After consultation, UKROFS adopted the EU standards with only three significant variations. These are that the UK standards maintain the requirement that meat animals to be sold as organic (except chicks) must be born and bred on an organic unit, and not brought in as conventional and converted; maintain specific UK standards in respect of BSE; and provide a derogation shorter than permitted for poultry producers to meet the flock sizes and stocking densities prescribed and limit the producers able to benefit from the derogation.[147] Concerns brought to our attention over the livestock regulations focussed both on these differences and on the EU regulations as a whole.

58. To look first at the broader picture, we received many representations, particularly from scientists and vets, that the regulations might have a negative impact on animal health in their restrictions on the use of allopathic remedies and antibiotics. This could have a significant effect on the number of organic farmers in the UK because potential recruits might feel that they are unable to comply with the regulations.[148] Particular problems arise with the treatment of parasites and sheep scab.[149] Both the Scottish Agricultural College and the NFU believed that the rules did not allow sufficiently for differences in geography, climate and farming practices across Europe.[150] The British Cattle Veterinary Association and others also raised questions about the efficacy and safety of homeopathic medicines which are allowed under the regulations.[151] UKROFS offered assurance that these fears are misplaced and affirmed that "there is sufficient flexibility to allow the standards to be operated effectively under most UK conditions".[152] It believed that the provisions of the regulations were "sufficient in most cases

---

[144]Q 513.
[145]Ev. p. 185, para 15.
[146]Ev. p. 101.
[147]Ev. p. 123, para 11.
[148]Q 64.
[149]Q 79.
[150]Ev. p. 11; Q 63; Q 237.
[151]Ev. p. 192, para 6.
[152]Ev. p. 252, para 1.5.

to ensure the health and welfare of organic stock".[153] **It is notable that their assurances were qualified in both cases. This is unsatisfactory.**

59. Turning to the regulations as operated differently in the UK, there are concerns, particularly within the poultry industry, that the decision to restrict the derogation to five years will be harmful in terms of competitiveness.[154] On the other hand, animal welfare groups, such as the RSPCA and CIWF, argued that the regulations did not go far enough.[155] UKROFS assured us that it had considered the competitiveness implications of the derogation agreed but had "concluded that the balance of advantage lay in a derogation which ... would encourage the industry to move relatively quickly to the standard set out in the Regulation."[156] The difficulty is that imports from EU competitors which do not meet these standards are perfectly legal and leave UK producers with higher costs competing against products produced to lower standards.

60. Most witnesses agreed that the EU livestock regulations are "moving in the right direction"[157] and that welfare problems were potential rather than inevitable as a result. However, it is clear that the impact of the standards will have to be closely monitored. When asked about this, UKROFS could not respond further than to offer the assurance that it "will be putting in place arrangements to monitor the effect of the standards by means of close consultation with the industry".[158] **We find this unsatisfactory and we are concerned by the lack of resources within UKROFS to conduct the necessary research into either animal welfare or the competitiveness impact of the regulations. We recommend that the Government ensure that the impact of the EU livestock regulations upon animal welfare and upon the competitiveness of the UK industry be monitored over the next decade with a view to recommending changes if necessary. We further recommend that UKROFS be charged to take into account the competitiveness implications of any proposed regulations and to publish the results of its analysis before agreeing on any changes to organic standards.** One other lesson to be drawn from the development of the organic livestock rules is that organic farming is not necessarily synonymous with the highest animal welfare standards. We do not dispute that organic farmers want to adopt welfare-friendly production standards but it is clear that this is another area in which more research is needed.

Processing standards

61. The EU livestock standards closed a significant gap in the European regulations for organic production but witnesses agreed that there remained much work to be done in the processing industry. At the moment, there are difficulties over the number of non-organic processing aids and additives allowed under the regulation.[159] There are also questions over the adequacy of the rules set down for complex processing, specifically where a plant makes both organic and conventional foods using a continuous process.[160] The experience of one organic sugar processor had led him to conclude that "the Commission and UKROFS have given processors an easier ride than farmers" by "adding to the list of approved processing aids and additives to help food manufacturers to get round their processing problems".[161]

62. These difficulties are likely to become more acute as more conventional manufacturers start to produce organic sidelines. Processing is clearly an area which will have to dealt with by the Commission in the same sort of comprehensive reform as in the livestock sector. However, MAFF had been given to understand that the Commission had "no immediate plans to draw up proposals in this area" and had too much work arising from the livestock standards to be able to take on any other matters.[162] In the absence of EU standards, it is possible to take independent

[153] *Ibid*, para 1.6.
[154] Ev. pp. 238-9.
[155] Ev. p. 244; Ev. p. 246.
[156] Ev. p. 252, para 3.1.
[157] Q 89.
[158] Ev. p. 253, para 4.1.
[159] Ev. p. 156, para 10.
[160] Ev. p. 156, para 8.
[161] *Ibid,* para 14.
[162] Ev. p. 151.

initiatives, such as supermarkets working with suppliers to reformulate ingredients[163] or insisting on labelling to ensure that consumers are aware of the non-organic ingredients permitted in organic foodstuffs.[164] However, for the sake of consumer confidence, it is essential that the rules on processing are clarified and, given the boom in production of processed organic foods, the need is becoming ever more pressing. The Government should not allow the Commission to fall behind developments in this way. **We recommend that the Government work in the Council of Ministers to present the Commission with a deadline by which to develop new standards for organic processing.**

The role of UKROFS

63. At the very end of our inquiry MAFF announced that UKROFS was to undergo its quinquennial review. The first stage of the review, to be completed within six to seven months, will look at the effectiveness of the current arrangements for discharging the national competent authority functions prescribed by the EU organic farming regulations and the continuing need for the discretionary functions currently exercised by UKROFS, including standards setting, direct certification of producers and advice to Ministers.[165] We have heard several points raised about the conduct and role of UKROFS which should be brought to the attention of the inquiry.

64. In the first place, many witnesses complained about the delays experienced both in clearing imports and in registering organic farms.[166] In some cases, these had involved perishable products.[167] This situation held back the development of the market and caused huge frustration to both farmers and the food industry.[168] UKROFS defended itself by pointing out that delays were not always "a one-sided problem" but accepted that difficulties arose because of the lack of money available to it.[169] All witnesses agreed that UKROFS was under-resourced in terms of staff and funding, particularly in view of the rapid expansion of the organic market. **The present Chairman admitted that when he took up his post he was "appalled at the level of resourcing and the pressure which was put both on the civil service secretariat and upon the board members of UKROFS by the sheer size of the workload".[170] This will have to be resolved.** However, there was less universal sympathy expressed with the performance of UKROFS in general. One witness told us that UKROFS "just seems to be a body that is sitting there and doing not a lot at the moment"[171], whilst others complained about its "total lack of commercial thought".[172] Even the Soil Association argued that "the standard setting committee did not have the capacity or the expertise to do the job that needed to be done in the development of standards".[173]

65. It is evident that this is a particularly appropriate time to be examining the role and performance of UKROFS. Its work has been affected by the adoption of the EU livestock standards and is further threatened by the increasing pressure for it to become merely the rubber stamp of standards agreed by the certification bodies. On another front, the United Kingdom Accreditation Service (UKAS) is making bold moves to take over its accreditation role.[174] UKROFS put up no defence when asked why this should not be permitted.[175] These changes could leave UKROFS with only its role of inspecting imports, which could be done just as efficiently by a dedicated organisation outside the current structure. We recognise that many of the complaints we have heard about the certification procedures and standards are directed at the entire regulatory process run by UKROFS. **We accept that at the moment UKROFS is not getting the support it needs from MAFF in terms of staff or funding. Nevertheless, we**

---

[163]Ev. p. 74, para 4.7.
[164]Q 521.
[165]MAFF News Release 384/00, 1 November 2000.
[166]e.g. Q 130.
[167]*Ibid.*
[168]Ev. p. 32; Ev. p. 78, section 5.5.
[169]Q 586.
[170]Q 547.
[171]Q 397.
[172]Ev. p. 95.
[173]Q 510.
[174]Ev. pp. 242-3.
[175]Q 561.

believe that there is scope for a complete reconsideration of its role. There is room for it
to acquire a higher profile, as was hinted at by the current Chairman's intervention in the
GM debate, and to perform a valuable role as the regulator between the certification
bodies and the Government, but it is clearly not fulfilling that potential at the moment. We
await the results of the review with great interest.

V. FINANCIAL ASSISTANCE FOR ORGANIC PRODUCTION

Public assistance

66. Financial support for organic production may be granted by the Governments of Member
States under European Union agri-environment measures (EC regulation 2078/92). All Member
States have taken advantage of this regulation to provide support for conversion and, with the
exception of the UK, for maintenance of organic production. Table 5 opposite sets out the
current rates applicable across the EU.

Table 5: A Comparison of Organic Aid Rates in the United Kingdom and Other EU Member States

EU country	Year of data	Payment for conversion UK£/hectare/year					Number of years paid	On-going payments	
		Arable area aid eligible land	Other crops or improved grassland	Unimproved grassland or rough grazing	Horticulture	Permanent crops		YES/NO	Payment UK£/hectare/year
United Kingdom[a]	2000	£90	£70	£10	£70	£90	5	NO	NIL
Austria	2000	£204	£74-£204	£39	£271	£497	5	YES	£39-£497
Belgium	1998	£126	£208	£207	£209	£586	2	YES	£78-£517
Denmark	1999	£38-£192	£38-£192	£38-£192	£38-£192	£38-£192	2	YES	£48-£192
Finland[b]	1998	£144-£246	£144-£315	not applicable	£315	£624	1-3	YES	£110-£590
France[c]	2000	£113	£113-£189	£66	£113	£190	2-3	3 regions	£29-£351
Germany	1999	£72-£125	£72-£125	£72-£125	£72-£125	£344	5	YES	£57-£501
Greece	2000	£106-£187	£106-£187	£106-£187	£106-£187	£274-£523	5	YES	£106-£523
Ireland[d]	2000	£113	£113	£113	£151	£151	2	YES	£35-£122
Italy	1996	£155	£259	£259	£259	£724	2	YES	£155-£724
Luxembourg	1999	£114	£114	£114	not applicable	not applicable	5	YES	no data available
Netherlands	1998-1999	£163	£62-£125	£62-£125	£813	£1,626	5	YES	£62-£125
Portugal	1998	£76-£252	£76-£252	£76-£252	£126-£252	£151-£504	2-3	YES	£126-£420
Spain	1996	£93-£117	£70-£117	£70	£187-£350	£163-£280	2	YES	£47-£233
Sweden	1998	£72-£177	£72-£177	£72-£177	£72-£177	£563	5	YES	£72-£563

(a) In Scotland, payments for improved and unimproved grassland are approximately £74 and £7 per hectare per year respectively. In certain cases in the UK, where the organic land is entered into another agri-environment scheme with land management conditions similar to aspects of organic farming, the aid rates are reduced. The reductions depend on the conditions involved; the lowest rates paid are £10/ha/year.

(b) Includes support provided under another agri-environment scheme in Finland.

(c) In France, continuing payments are made in the Centre, Nord Pas de Calais and Ile de France regions only, out of a total of 22 regions

(d) The range for continuing payments in Ireland includes an additional £75/ha/year for specialist growers with more than 1 hectare.

Source: SA, Ev. p. 106.

67. Within the UK, support was first offered under the Organic Aid Scheme (OAS) which offered flat rates for all land types and the lowest rates available in any of the Member States.[176] Between 1994 and 1999 just 400 participants entered the scheme in England, and we had cause to draw attention in our First Report of Session 1998-99 to the underspend of some £550,000 in payments to individuals in 1997-98.[177] The scheme was reviewed in April 1998 and relaunched a year later as the Organic Farming Scheme (OFS) with "considerably" higher rates of aid and a "greatly increased" overall budget of £11.35 million in 1999-2000 and £12 million in 2000-01, compared with a high under the OAS of £1 million in 1998-99.[178] The rates of aid were also made variable, ranging from £450 per hectare for land eligible for AAPS payments, £350 per hectare for other improved land and £50 per hectare for unimproved land, paid over a five year period.[179] The payments cover the period of conversion. There are no payments aimed at maintenance of organic farming. Special provision was made when the scheme was announced for those who had joined the OAS on or after 2 April to switch to the new OFS after a year.

68. Take-up of the OFS was extremely high, with the scheme attracting 1,214 applications. Consequently, within four months the scheme was closed to new applications and the Government issued a consultation document on its operation. The results of this consultation were announced in November 2000 and it is now expected that the scheme will reopen in January 2001 for first payments in April, with a few minor changes such as extending the period for applying for aid after registration from three months to six months.[180] The budget for organic conversion money has also swelled with the implementation of the England Rural Development Programme (ERDP) which provides for expenditure of £139 million to 2006. Table 6 below illustrates total Government funding for organic conversion in the UK under the OAS and OFS, projected to the end of the ERDP. A further review of the OFS is planned for 2003 as part of the general review of the ERDP and Mr Morley assured us that it was possible that the budget might be adjusted again at that point.[181]

Table 6: OAS and OFS funding (£'000)

	Organic Aid Scheme	Organic Farming Scheme
1995-96	261	
1996-97	374	
1997-98	571	
1998-99	1,026	
1999-2000		12,037
2000*		13,400
2001		20,600
2002		19,000
2003		19,900
2004		22,300
2005		22,900
2006		22,600

* year 2000 onwards years from 16 October to 15 October
Source: HC Deb., 10 Nov. 2000, Cols 409W–410W; 29 Nov. 2000, Cols 661W–662W.

---

[176]Ev. p. 34.

[177]First Report of the Agriculture Committee, Session 1998-99, *MAFF/IB Report 1998 and the Comprehensive Spending Review*, HC 125, para 20.

[178]Ev. p. 134, paras 22-23.

[179]*Ibid*, para 22.

[180]Q 603; MAFF News Release 386/00, 2 November 2000.

[181]Q 603.

69. Mr Morley summarised the response to the consultation on the OFS as a feeling "that the current structure of the scheme was about right".[182] In contrast, we heard a number of complaints about the scheme, particularly in comparison to those operating in other Member States. In general, it was agreed that the rates payable were acceptable[183] but a number of witnesses argued that they were inadequate to meet the needs of specific sectors, especially horticulture where the costs of conversion were variously estimated as typically £2-3,000 per hectare or £10,000 per hectare over a three year period.[184] A case was also made that the UK should provide maintenance payments.[185] However, the strongest criticisms were of the stop-go approach taken to funding, which was seen to have limited the development in the UK organic sector at a critical time. This affected processors and retailers, as well as farmers who were trying to make crucial decisions whilst uncertain whether support would be available and if so, at what level. There is a constant complaint from farmers that they are forced to phase their conversion in order to fit into periods when funding is available and it is clear that some farmers have lost out because of the unfairness and lack of information. As with the other issues raised, this had severe implications for the competitiveness of the UK industry compared to the industries of countries where funds were continuously available.[186] It also created difficulties for the certification bodies who had to deal with a huge increase of applications at the beginning of the year which could have been handled more easily if spread across the whole twelve months. **Given that there is a programme, we believe that the disruption in the provision of aid for organic farming at this crucial time has been highly regrettable. The Government should seek to ensure that the OFS is administered to provide even funding and applications across the whole year.**

Private assistance

70. The consumer demand for organic products and the perceived environmental benefits offered by organic farming have led some private sector organisations to give financial support to producers. Examples in the retail sector given to the Committee included £3 million from ASDA over the next three years to be spent on encouraging its meat suppliers to convert to organic farming methods.[187] Wessex Water Company is also introducing a scheme of top-up payments of £40 per hectare to farmers in recognition of the contribution organic farming can make to reducing pollution of water by pesticides.[188] We understand that "at least two other water authorities" are examining the feasibility of such schemes.[189] Private sector assistance is thus available and increasing and it is to be encouraged as a means of risk-sharing but we accept that it is unlikely that the private sector, whether in the food industry or elsewhere, will be able or willing to offer sufficient incentive, either in terms of the number of schemes or the rates of payments, to make a real difference in the number of farmers converting to organic production. If the Government wishes to see an increase in organic farming in the UK, public subsidy will achieve a greater rate of conversion than otherwise would be the case. **However, the Government should discuss the design of its subsidy regime with retailers, processors and the water industry. It should encourage OFWAT to review whether water companies should be obliged to offer top-up payments.**

Rationale for Government assistance

71. **Before determining how best to offer financial assistance to organic producers, it is essential that the Government be clear as to the rationale for doing so and the objectives it wishes to achieve through this expenditure. These objectives must be tightly defined and made public.** There is, of course, a perfectly reasonable case against any aid for conversion. After all, if the market really is expanding, if retailers are anxious to provide more organic products and if organic produce commands a premium, why should the state intervene? The

[182]*Ibid.*

[183]Ev. p. 200; Ev. p. 19, para 17.

[184]Q 137: Ev. p. 34.

[185]Ev. p. 165, para 10.6.

[186]Ev. p. 19, para 18.

[187]Ev. p. 224.

[188]Ev. p. 35; Q 487.

[189]Q 489.

only argument to do so is that other countries assist organic production and that the UK has a broad economic interest in trying to satisfy as much demand as possible from domestic sources. More specifically, this argument can be broken down into three factors which can be put forward to justify public support: the gap between supply and demand, competitiveness and the pursuit of specific public goods.

72. First, on the gap between supply and demand, we recognise that the growth in the organic sector has been largely market-driven but that the costs of conversion are such that they deter most would-be organic farmers from taking the step without some form of subsidy. Some farmers have converted without public assistance but the evidence is clear that the numbers slumped when Government money was not available.[190] Of course, this also has its down-side: support could lead to a distortion of the market and might be harmful to those who are already operating in the industry.[191] In the prevailing economic circumstances, it is also possible that subsidies will attract farmers who will not be able to follow through the conversion of their land to organic production and who are motivated by the hope of finding a way out of existing problems rather than by a well thought out vision of sustainable organic production. However, the yawning gap between demand and supply, even in products capable of being grown in the UK, indicates that there are market opportunities here which, as we have seen, may not be wholly addressed by the private sector.

73. Second, there is the question of the competitiveness of the UK industry, extending beyond primary production to processing and marketing. When schemes in other Member States, in particular, are more generous, UK farmers are disadvantaged both in terms of their ability to offer organic goods and in terms of the cost of production and hence the price of their produce. Mr Morley accepted that the Government "should look at what other European countries are doing in relation to their support regimes" because "It has an impact on competitiveness".[192] This is also the case where schemes differ within the United Kingdom. For example, more financial assistance has recently been made available in Wales. Clearly, the National Assembly of Wales has the absolute right to take such action but it leaves a disparity of funding between the territories of the UK which inevitably has implications for competitiveness.

74. Third, there is justification for supporting organic farming with public money if it contributes towards wider benefits which the Government wishes to achieve. Most of those arguing for further Government assistance did so on the basis that organic farming provides public goods beyond the immediate effect on the farmer. The Soil Association claimed that large amounts of public expenditure could be saved through expansion of organic farming[193] whilst Dr Lampkin listed many public goods in terms of social, environment, animal welfare, rural development, food quality and health issues.[194] Of course, many of these claims are disputed and we also heard opposing arguments over supporting organic farming *per se*. Professor Sir John Marsh advised us that "Rewarding a particular system, which is defined in terms of an inflexible system of rules, is not helpful".[195] He believed that "Policy will make a much more positive impact if it identifies and rewards the specific outcomes it seeks."[196]

75. In written evidence MAFF was anxious to assure us that Government support for organic farming was predicated not upon the more doubtful claims made for public goods but on the contribution it could make to the wider Government aims of a "prosperous, forward-looking and sustainable" farming sector, which was "competitive, and flexible enough to respond quickly and effectively to market changes and consumer needs".[197] The exception to this was that Government support "takes account of evidence that the organic system of farming leads to certain environmental benefits".[198] In oral evidence, Mr Morley clarified that he believed that "there are a range of benefits, economic, social and environmental, that organic farming brings"

---

[190]Ev. p. 175.
[191]Ev. p. 172, para 6.4.
[192]Q 700.
[193]Ev. p. 96, section 1.4.
[194]Ev. p. 20, para 30.
[195]Ev. p. 3.
[196]*Ibid.*
[197]Ev. p. 131, para 2.
[198]Ev. p. 131, para 3.

and that Government support was justified for these reasons, as well as for issues of consumer choice and import substitution.[199] He emphasised that "there are benefits, particularly environmental benefits, which organic farming gives, which we do recognise as a public good."[200] The question remains whether these same benefits could be achieved by other means at a lower cost to the public purse. **We believe that the benefits to be secured by organic farming need to be far more closely defined so that the Government can set measurable and achievable objectives for its financial assistance to organic farming.** In the absence of other evidence, we recognise that the greatest claims that can be made for organic farming are environmental benefits. We believe that the OFS needs to be more closely focussed on these benefits so that farmers are paid for delivering recognisable public goods, rather than merely for converting to organic farming as a good in itself. It also means that aid for organic farming could eventually be assimilated to other schemes based on payment in return for specific outcomes or practices like, for example, countryside stewardship. We examine below suggestions for how this could work.

Options for consideration in the 2003 review

76. The review of the ERDP in 2003 will give the Government a useful opportunity to consider the objectives and structure of its support for organic farming. In discussing criticisms of the existing scheme, we have already touched on some options which should be considered. These include differential rates of aid to reflect the true cost of conversion in sectors such as horticulture, although we recognise that research remains to be done on establishing these costs. **We recommend that in advance of the review the Government commission such research in order that its consideration of differential payments be properly informed.** Another suggestion is that the Government should target aid at sectors which are lagging behind the general trend towards organic conversion. Apart from horticulture, the obvious example here is arable land which has implications not just for processed foods but for animal feeds and hence the livestock industry. Yeo Valley Organic Company put forward an imaginative scheme for encouraging arable conversion through the use of set-aside payments.[201] **Such targeting, supported also by the NFU, should be included in the options for consideration, if the current trend continues.**

77. A far more radical approach, which met with near universal approval, was the restructuring of organic farming support within an organic stewardship scheme. This would recognise that the payments are made to farmers for the benefits they bring to the environment and would give the Government a means by which they can specify the benefits to be realised in each case. It would meet the demand for a maintenance scheme in that the money would be available to farmers after conversion as well, including those who at the moment are disqualified from applying for OFS because they have received organic funding in the past or had registered organic land before the OAS/OFS started.[202] The Soil Association was strongly of the opinion that "an organic stewardship approach is a better way of doing it than a front-end loaded scheme which gives a carrot tomorrow and really does not encourage people to think through their marketing and how they are going to deal with their business over time".[203] The Organic Farmers and Growers also preferred this approach which would ensure a steady growth in the market, rather than a sudden surge as happened at the moment.[204] Finally, a stewardship scheme would represent a move away from production payments, as we have consistently advocated should be the trend for all farming subsidies.

78. Mr Morley was sympathetic to the principle of an organic stewardship scheme,[205] although he stressed that "at the moment our priority is to provide funds for conversion".[206] More recently, the Minister, the Rt Hon Nicholas Brown MP, has expressed stronger support for such a scheme in his address to an organic farming conference. There are many details which

---

[199]Q 594.
[200]Q 600.
[201]Ev. p. 62.
[202]Ev. p. 249.
[203]Q 542.
[204]Q 543.
[205]Q 597.
[206]Q 611.

would have to be worked out before the proposal could be put out for consultation. For example, whether the scheme should be a specifically organic stewardship scheme or whether there should just be greater recognition that organic farming may be intrinsically better positioned to meet the criteria of ongoing agri-environmental programmes, such as countryside stewardship, through its delivery of environmental benefits. We believe that both options should be explored. There may also be a continuing need for subsidies to cover the initial costs of conversion over and above the level of ongoing payments in certain sectors. Another factor to be considered is the current "profit-forgone" structure of agri-environmental support which might need to be revisited. There are two features, however, which we would wish to see included in any scheme. First, **we recommend that applicants to the scheme be required to produce a business plan which is accompanied by a statement as to its validity from a qualified adviser, such as a bank, accountant, consultancy or agricultural organisation.** This suggestion was rejected in the Government's recent review as too bureaucratic[207] but we believe it is essential – the discipline of the exercise also has value. Second, **we recommend that, whatever scheme is devised, it be flexible, locally-run and as unbureaucratic as possible.** We recognise that Governments seek administrative simplicity but if the change in regional representation of MAFF, with its greater emphasis on using modern information technology flexibly, is to mean anything, it should be possible to manage a flexible scheme responsive to local conditions without turning it into a bureaucratic nightmare. The farmers who gave evidence to us asked for simplicity, certainty, fairness and predictability.[208] These should be the hallmarks of all good administration and we expect attention to be paid to these factors in this case. **We recommend that the Government devise proposals for an organic stewardship scheme as the centrepiece of its review of organic farming support in 2003, taking into account the need for clearly defined goals and for flexibility, simplicity and predictability. These proposals should be accompanied by a statement of objectives and plans for the achievement of those objectives, including the resources to be allocated to their achievement.**

VI. OTHER AREAS OF GOVERNMENT POLICY

Government support for organic farming

79. Although financial subsidies to organic farmers are the most obvious and debated aspect of public sector support for the sector, there are other ways in which the Government offers assistance. MAFF included in its summary of such measures: responsibility for setting and maintaining standards through UKROFS and negotiating organic standards within the European Union.[209] We have discussed these matters in Section IV above. In this section of the Report we concentrate on two further measures highlighted by MAFF, namely, research and development and advice and training, and on a further potential role, proposed by many witnesses, of devising a Government strategy for the expansion of the organic sector.

Research and development

80. The policy objectives of MAFF's research programme for organic farming are: to remove constraints to organic production in order to make organic conversion more attractive to conventional farmers; to provide information on the economics of organic conversion to inform the grant scheme; and to provide information on the environmental impact of organic farming in comparison with conventional farming, to inform the main policy rationale for organic support.[210] Within the programme, the largest element (40 per cent) consists of system studies of organic production within particular sectors (dairy, beef and sheep, arable, field vegetables, pigs, poultry). Generally associated with linked commercial organic farms, the studies cover assessment of farming techniques and system economics. The rest of MAFF's research is into economics (the causes of differences in profitability between organic farms and a general assessment of the economics of organic farming); specific constraints to organic production (control of pests, diseases and weeds, the management of animal health, the supply of nutrients,

---

[207]Q 620.
[208]Q 335.
[209]Ev. p. 131, para 4.
[210]MAFF Research Strategy 2001-2005 Consultation Document, Chief Scientist's Group, August 2000, p. 67.

the use of break crops, the selection of appropriate varieties, and the production of organic seeds); impacts of organic farming upon the environment (including biodiversity implications and the impact of organic farming on soil health); and technology transfer (a strong emphasis in each project on information dissemination, reviews of research and the collation of advisory information).[211]

81. The research programme is informed by advice from the UKROFS Committee on R&D which is intended to reflect the research priorities of the organic farming community. There is also "good contact" with other research bodies with expertise in organic research within the UK, other funders of such research and the organic sector bodies.[212] The programme will be reviewed in 2001 with input from these bodies. The Chief Scientist's Group expected that the share of resources allocated to environmental interactions would increase in the future in order to demonstrate the environmental benefits of organic farming upon which government support for the sector is predicated.[213] Other priorities identified included the ongoing research into technology transfer and constraints on production and additional areas including "sensitivity of economic performance to price premia and the effect of subsidies; horticultural conversion and production; [and] other standards issues, particularly in relation to the recently agreed Community livestock rules, and possibly further work on the interaction of organic farming with GM crops." [214]

82. Issues raised by witnesses concerning MAFF's research into organic farming centre on the size of the programme as well as its scope. Forecast expenditure for 2000-01 is £2.1 million, with a total of £5.8 million over the lifetime of the current projects.[215] This is out of an overall MAFF spend on research of approximately £104 million.[216] The percentage spend on organic research reflects the percentage of land currently farmed organically at 2.1 per cent and 2.3% per cent, respectively. It has been argued that in order to help organic farming expand, the budget for research and development needs to grow significantly: the Soil Association put the case for 30 per cent of the overall budget.[217] At the moment, there are no plans for an increase on anything like this scale. The Consultation Document on MAFF's research 2001-2005 indicated that "The size of the research programme [for organics] is likely to remain at about the current level for the period of this Strategy".[218] This would imply that the relative proportion of the research budget spent on organics is going to fall increasingly behind the proportion of organic production in the UK. In considering the case for faster growth, we are mindful that much of MAFF's "conventional" research is highly relevant to the organic sector (and *vice versa* much organic research has relevance to conventional production). For example, Mr Morley drew our attention to MAFF's £8 million research programme on biological control.[219] In addition, other funding has been made available in the form of a Government grant of £2.2 million to establish a European organic top fruit centre at East Malling, under the auspices of Horticultural Research International.[220] **Nevertheless, we believe that the Government should consider increasing its budget for organic research and development to take account of its expectations for the market and in line with the need for further research into the areas we outline below.**

83. We identified four broad areas into which more research was necessary in order to encourage successful organic production in the UK. These are environmental impacts; technical research, including pesticides; animal welfare issues; and the scientific basis of the claims made for organic farming and the protocols imposed on organic farmers. As we have noted above, MAFF's research programme includes work on most of these elements but it is worth highlighting the apparent gaps or shortcomings in that programme. With regard to the environment, we were told that this was the area in which most work had been carried out (for

[211]*Ibid*, p. 68.
[212]*Ibid*, p. 67.
[213]*Ibid*, p. 69.
[214]*Ibid*, p. 69.
[215]Ev. p. 134, para 27.
[216]MAFF Research Strategy 2001-2005, Annex 2.
[217]Ev. p. 103.
[218]MAFF Research Strategy 2001-2005, p. 69.
[219]Q 629.
[220]*Ibid*.

example, the studies by English Nature and the Soil Association into the biodiversity benefits of organic farming)[221] but that it was still an issue which needed exploring in more depth, particularly with a view to identifying the components within organic farming which deliver environmental benefits.[222] On the other hand, HRI complained that MAFF's organic R&D budget was "heavily focussed towards environmental impacts", leaving insufficient "scope for strategic, production-targeted R&D aimed at improving yields, quality, efficacy of disease control and overall efficiency" as is necessary to meet the Government's policy of increasing organic production in the UK.[223] The Soil Association also believed that "We desperately need more research into some techniques within organic farming".[224] We were told that there were particular problems with the development and registration of organic pesticides because of the size of the market relative to the costs of registration.[225] A case could therefore be made for the Government to step in.

84. The other two areas identified as lacking in research relate more to doubts over the benign effects of organic farming. Animal welfare has come to the fore with the agreement of the livestock regulations, with the result that there have been calls for much more research into organic farming practices in this sector and possible improvements. The Veterinary Department at the University of Reading submitted a list of projects which should be undertaken and the Royal College of Veterinary Surgeons drew our attention to the particular difficulties of alternative medicines which have not been tested to the same rigorous standards of safety and efficacy as conventional ones.[226] The National Office of Animal Health (NOAH) wanted to extend this to include "more research ... devoted to an open-minded investigation of some of the claims made by organic farming".[227] This call was seconded by other witnesses. Not surprisingly, most of them were not what might be termed part of the organic movement but not all of them were hoping to see the case proven against organics. For example, HRI, which runs a considerable programme of organic research in addition to the new European Centre for organic fruit,[228] was one of the strongest proponents of the "need to understand and underpin systems and protocols".[229] Such research could provide the scientific basis for organic agriculture, the absence of which was decried by another expert, Professor McKelvey.[230] We are aware that some work has been carried out into the claims of the organic movement, most recently by the Royal Agricultural Society of England,[231] but **we believe that there are three reasons why such research is essential and should be carried out by a reliable source, independent of either the conventional or the organic sector. First, it is important that Government policy be based on hard fact, rather than supposition. Second, it would assist the organic sector if it were known that there was a scientific basis for the demands they were making of their producers in setting standards and the promises they were offering to consumers. Third, such research should also isolate the elements within organic production protocols which lead to the desired benefits, with the result that these techniques may be applied more effectively both on conventional and organic farms. We recommend that MAFF commission additional research into the environmental implications, technical issues, animal welfare and verification of claims made in connection with organic farming on public policy issues such as food safety to supplement its existing programme.**

85. Research on its own is not enough and we welcome the recognition of MAFF that it is equally important to disseminate the results of research. **We stress the benefits of treating organic and conventional production as part of the same spectrum, with the outcome of research in one sector being applied to the other.** We should also like to draw attention to experiments in integrated crop management, such as those run by the CWS in Leicestershire

---

[221]Ev. pp. 228-34, annex 2; SA (2000), *The Biodiversity Benefits of Organic Farming.*
[222]Ev. p. 164, para 10.2.
[223]Ev. p. 187, para 27.
[224]Q 532.
[225]Q 214.
[226]Ev. pp. 203-204, para 6; Ev. p. 245.
[227]Ev. p. 215.
[228]Ev. p. 184, para 5.
[229]Ev. p. 186, para 26.
[230]e.g. Q 89.
[231]*Shades of Green: A review of UK farming systems*, RASE, November 2000.

which seek to identify and apply the best practices from both systems. We believe that there is much to be learned from such experiments. A further consideration is collaboration with other institutes to ensure that efforts are not duplicated, either within the UK or more widely. HRI pointed to the existence of "well funded research establishments with programmes dedicated to organic research and production" in several other European countries.[232] We expect MAFF to make full use of the opportunities for co-working with these institutes. However, we recognise that there is also an issue of competitiveness and it is vital that the UK maintains a scientific edge in this expanding area in order to boost UK industry. The forthcoming review of MAFF's research programme is an ideal opportunity for MAFF to respond to the concerns expressed about the scope of its research and development of organics. We will examine the results with close interest.

Advice and training

86. One of the farmers to whom we spoke in the course of the inquiry told us that, in undergoing the conversion of his farm, "I think a very practical advisory service on agronomy would be a very valuable thing to me, perhaps more so than the financial side of it".[233] Advice is available to farmers considering conversion from the MAFF-funded Organic Conversion Information Service (OCIS). OCIS provides a dedicated telephone helpline (run by the Soil Association) and a free advisory visit of up to one and a half days from advisers based at the Organic Advisory Service at Elm Farm Research Centre to help the farmer decide whether conversion is a viable proposition.[234] MAFF has also commissioned an organic research database and software to assist advisors, the results of which were expected by the end of 2000.[235] Training for farmers is funded within the new rates of payment for the Organic Farming Scheme and could be increased if the Government receive suitable proposals for organic training schemes under the ERDP budget for training for rural sectors.[236]

87. Mr Morley was satisfied with the Government's strategy for advice and training[237] but others felt that, while the quality of advice had improved,[238] there was still an issue to be resolved around the provision of training and advice services not just for farmers considering conversion but for existing producers, processors, retailers, consumers and other groups working with the agricultural sector, such as vets.[239] We note the pressures placed on OCIS with the huge increase in the number of farmers wanting initial advice about conversion. However, it is clear that there is a need for practical, readily available advice for farmers past this stage who are no longer eligible for OCIS services. Such advice need not be provided directly by MAFF, as indeed the OCIS helpline is contracted out at the moment, but it probably does require MAFF funding and planning to ensure that the level of service meets the demands made upon it. For example, one farmer praised the Lincolnshire Organic Producers Limited for its constructive advice, an organisation subsidised by MAFF but, we are told, unique of its kind.[240] **We recommend that MAFF review the provision of advice to the organic sector in the light of its commitment to organic farming, to ensure that the advice available is adequate and meets the needs of producers in conversion and post conversion and others involved in the sector.**

88. Increasing advice services will require a corresponding increase in the number of trained advisers. At the moment, the provision of education is patchy. We understand that there is only one MSc course in the UK on organic farming, which is run by the Scottish Agricultural College.[241] There are examples of good practice. For instance, the SAC is developing a distance learning package for organic farming and runs training days for farmers and others involved in

---

[232]Ev. p. 187, para 36.
[233]Q 269.
[234]Ev. p. 134, para 26.
[235]Ev. p. 134, para 28.
[236]Q 596.
[237]Q 694.
[238]Q 261.
[239]Ev. p. 21, para 31.
[240]Q 265.
[241]Q 53.

the sector, such as seed merchants and vets.[242] In Wales the Organic Farming Centre is publicly funded by the National Assembly and has a remit to co-ordinate the dissemination of information including R&D, advice, training, education and demonstration farms.[243] Similarly, the Elm Farm Research Centre in England is to be commended. However, more needs to be done to bring these initiatives together to provide high quality training across the board. **We recommend that the Minister actively encourage the development of organic training schemes within the English Rural Development Programme and promote the development of training schemes in the UK.**

Government targets and strategy

89. The Prime Minister has said that Government plans envisage a trebling of the area under organic farming in the UK by 2006.[244] This would amount to around 6 per cent of agriculturally used land. There is, however, a concerted campaign to tie the Government to much more ambitious targets. The Organic Food and Farming Targets Bill, a Private Members Bill introduced in Session 1999-2000, aimed to ensure that 30 per cent of agricultural land will be organic by 2010 and that 20 per cent of the food consumed will be organic by that date.[245] It drew on the experience of other European countries which have set targets for the growth of their organic industries, such as Sweden, Austria, Denmark and Finland, although not all of the targets have been met.[246] The campaign behind the bill is steered by organic organisations and others and supported by a wide range of organisations, from supermarkets to statutory agencies, environmental groups and trade unions. Supporters argue that the advantage of setting targets linked to a long-term Government strategy would be to "help the sector develop smoothly" and "give the confidence to growers, farmers, retailers and investors that the organic sector is set on a course of growth".[247] Not all organic bodies are in favour, however, with the Organic Farmers and Growers Ltd telling us that it wanted no targets "whatsoever".[248]

90. The bill is unlikely to succeed at Westminster, although it will be taken up and re-presented to the House of Commons in the current Session. In any case, witnesses who supported the campaign generally agreed that the purpose of the bill was to raise awareness, rather than to prescribe an exact target.[249] A further purpose of the bill was to persuade the Government to adopt an Action Plan. It is pleasing that the Minister for Agriculture has now accepted this, following his recent speech at the Circencester conference of the Soil Association. The Soil Association recognised that the real need was "to plan for growth, to put in place the structures which we need to enable that growth to happen sensibly, for us to ensure that the market can be developed at the right rate as we develop the production base".[250] In written evidence, the Association argued that "the Government urgently needs to adopt a long-term strategy for the development of the organic sector".[251]

91. Mr Morley countered by arguing against a fixed strategy and in favour of flexibility.[252] He believed that while the Government did "have to try and look ahead and try and project trends, we really feel that, at the moment, the organic sector is being market driven by market demand, and we think that is quite right and proper".[253] This position could alter if premia for organic produce disappeared and the Government wished to retain the benefits it perceived organic farming as supplying.[254] **We accept Mr Morley's point that the inclusion of organic farming within the ERDP gives some flexibility to respond to developments in the sector as far as the budget is concerned. We are not in favour of a _dirigiste_ approach to**

---

[242]Q 53.

[243]Ev. p. 21, para 33.

[244]Speech to the NFU, 1 February 2000.

[245]Ev. p. 174, section 2

[246]Ev. pp. 175-6, section 4.4.

[247]Ev. p. 174.

[248]Q 544.

[249]Q 469; Q 461; Q 545.

[250]Q 545.

[251]Ev. p. 103, section 7.1.

[252]Q 600.

[253]Q 600.

[254]Q 597.

agriculture in the UK. Agriculture must respond to the market-place and farmers need to adopt clear plans that will allow it to do so. This is particularly true of the organic sector. However, we believe that the Government has a role in analysing the organic supply chain for bottlenecks and imbalances and devising policy tools to help remedy these. There is merit in the Government setting out long-term projections of the money available for conversion and for assistance to address supply chain difficulties.

VII. CONCLUSIONS AND SUMMARY OF RECOMMENDATIONS

92. Our principal conclusions and recommendations are as follows:

Claims made for organics

1. We have seen no evidence to enable us to state unequivocally that any of the many claims made for organics are always and invariably true. All claims need to be properly evaluated in order to help consumers make their own judgements on the benefits of organic produce (paragraph 5).

2. We believe it important that the claims can be tested and verified in order that consumers know what they are really buying (paragraph 6).

Market for organics

3. It is clear that there is a huge opportunity for UK producers to expand still further into organic farming to meet a ready market (paragraph 22).

4. It is vital that the organic industry develops its ability to market its products effectively so that they appeal not to sentiment but to proven benefits. The industry may need to be less messianic and more marketing-orientated in its public presentations (paragraph 32).

Local marketing schemes

5. We recommend that the Government encourage the further development of local marketing schemes, such as farmers' markets and box schemes, through the provision of advice and ERDP funding (paragraph 33).

Supermarkets and organics

6. Supermarkets will be the main, although not the only, distribution channel for organic produce. It is critical that they are involved in the design of and encouraged to co-fund future initiatives to further organic conversion (paragraph 37).

Supplier partnerships

7. We recommend that the Government work with the bodies responsible for the promotion of organic production to ensure that rural development funds are channelled into the development of supplier partnerships and farmer-controlled co-operatives in the organic sector (paragraph 38).

Abattoirs

8. We welcome the additional aid for small and medium sized abattoirs announced in the Rural White Paper and await with interest details of the package and we urge the Government to stimulate the development of new small abattoirs, including mobile abattoirs (paragraph 40).

Certification bodies

9. **The multiplicity of bodies with their different standards and symbols is a significant weakness and we believe that the certification bodies should be encouraged by the Government in their efforts at closer co-operation, which may lead ultimately to mergers (paragraph 46).**

Farm assurance schemes and organic certification

10. **We recommend that MAFF facilitate discussions between the farm assurance schemes and the organic certification sector with a view to ensuring agreement on common core values and inspection protocols and with the goal of a single inspection process and shared symbols (paragraph 48).**

European regulation on organic production

11. **We recommend that the Government ensure that the European Commission reports regularly on the implementation of the regulation and actively encourage the European Parliament to monitor this implementation. The Government should produce a "Non Paper" for distribution at the Agriculture Council to further this end (paragraph 52).**

12. **We further recommend that MAFF be pro-active in drafting EU regulations and ensuring their scientific validity before they are written into law. MAFF should also, either directly or through UKROFS or the FSA, seek to monitor the effect of regulations to ensure that other public policy objectives are not compromised (paragraph 52).**

Organic production standards in third countries

13. **Unless these discrepancies are removed, there is a real danger that confidence in organic food may be damaged (paragraph 53).**

14. **We believe that IFOAM accreditation has much to offer in gaining acceptance for the standards met by imports from third countries and that the Government should support its widespread adoption (paragraph 56).**

Setting standards

15. **We recommend that the Government endorse the involvement of the certification bodies in setting standards, with UKROFS acting as a check and balance in the system, and that the Government provide sufficient funding to ensure the rigour of standard-setting procedures (paragraph 56).**

EU livestock standards

16. **It is notable that UKROFS' assurances on the suitability of the EU livestock rules for the UK and their potential impact upon animal health and welfare were qualified in both cases. This is unsatisfactory (paragraph 58).**

17. **We find UKROFS' arrangements for monitoring the effect of the standards unsatisfactory and we are concerned by the lack of resources within UKROFS to conduct the necessary research into either animal welfare or the competitiveness impact of the regulations. We recommend that the Government ensure that the impact of the EU livestock regulations upon animal welfare and upon the competitiveness of the UK industry be monitored over the next decade with a view to recommending changes if necessary (paragraph 60).**

Competitiveness and standards

18. **We recommend that UKROFS be charged to take into account the competitiveness implications of any proposed regulations and to publish the results of its analysis before agreeing on any changes to organic standards (paragraph 60).**

Processing standards

19. **We recommend that the Government work in the Council of Ministers to present the Commission with a deadline by which to develop new standards for organic processing (paragraph 62).**

UKROFS

20. **The present Chairman of UKROFS admitted that when he took up his post he was "appalled at the level of resourcing and the pressure which was put both on the civil service secretariat and upon the board members of UKROFS by the sheer size of the workload". This will have to be resolved (paragraph 64).**

21. **We accept that at the moment UKROFS is not getting the support it needs from MAFF in terms of staff or funding. Nevertheless, we believe that there is scope for a complete reconsideration of its role. There is room for it to acquire a higher profile, as was hinted at by the current Chairman's intervention in the GM debate, and to perform a valuable role as the regulator between the certification bodies and the Government, but it is clearly not fulfilling that potential at the moment. We await the results of the review with great interest (paragraph 65).**

Organic Farming Scheme

22. **Given that there is a programme, we believe that the disruption in the provision of aid for organic farming at this crucial time has been highly regrettable. The Government should seek to ensure that the OFS is administered to provide even funding and applications across the whole year (paragraph 69).**

The private sector and organic subsidies

23. **The Government should discuss the design of its subsidy regime with retailers, processors and the water industry. It should encourage OFWAT to review whether water companies should be obliged to offer top-up payments (paragraph 70).**

Objectives of Government assistance for organic farming

24. **Before determining how best to offer financial assistance to organic producers, it is essential that the Government be clear as to the rationale for doing so and the objectives it wishes to achieve through this expenditure. These objectives must be tightly defined and made public (paragraph 71).**

25. **We believe that the benefits to be secured by organic farming need to be far more closely defined so that the Government can set measurable and achievable objectives for its financial assistance to organic farming (paragraph 75).**

The 2003 review

26. **We recommend that in advance of the review the Government commission research into the cost of conversion in different sectors in order that its consideration of differential payments be properly informed (paragraph 76).**

27. **The targeting of aid at sectors which are lagging behind the general trend towards organic conversion should be included in the options for consideration, if the current trend continues (paragraph 76).**

28. **We recommend that applicants to any organic subsidy scheme be required to produce a business plan which is accompanied by a statement as to its validity from a qualified adviser, such as a bank, accountant, consultancy or agricultural organisation (paragraph 78).**

29. **We recommend that, whatever scheme is devised, it be flexible, locally-run and as unbureaucratic as possible (paragraph 78).**

30. **We recommend that the Government devise proposals for an organic stewardship scheme as the centrepiece of its review of organic farming support in 2003, taking into account the need for clearly defined goals and for flexibility, simplicity and predictability. These proposals should be accompanied by a statement of objectives and plans for the achievement of those objectives, including the resources to be allocated to their achievement (paragraph 78).**

Research and development

31. **We believe that the Government should consider increasing its budget for organic research and development to take account of its expectations for the market and in line with the need for further research into the areas we outline below (paragraph 82).**

32. **We believe that there are three reasons why research into the claims made for organic farming is essential and should be carried out by a reliable source, independent of either the conventional or the organic sector. First, it is important that Government policy be based on hard fact, rather than supposition. Second, it would assist the organic sector if it were known that there was a scientific basis for the demands they were making of their producers in setting standards and the promises they were offering to consumers. Third, such research should also isolate the elements within organic production protocols which lead to the desired benefits, with the result that these techniques may be applied more effectively both on conventional and organic farms (paragraph 84).**

33. **We recommend that MAFF commission additional research into the environmental implications, technical issues, animal welfare and verification of claims made in connection with organic farming on public policy issues such as food safety to supplement its existing programme (paragraph 84).**

34. **We stress the benefits of treating organic and conventional production as part of the same spectrum, with the outcome of research in one sector being applied to the other (paragraph 85).**

Advice and training

35. **We recommend that MAFF review the provision of advice to the organic sector in the light of its commitment to organic farming, to ensure that the advice available is adequate and meets the needs of producers in conversion and post conversion and others involved in the sector (paragraph 87).**

36. **We recommend that the Minister actively encourage the development of organic training schemes within the English Rural Development Programme and promote the development of training schemes in the UK (paragraph 88).**

Government targets and strategy

37. **We accept Mr Morley's point that the inclusion of organic farming within the ERDP gives some flexibility to respond to developments in the sector as far as the budget is concerned. We are not in favour of a *dirigiste* approach to agriculture in the UK. Agriculture must respond to the market-place and farmers need to adopt clear plans that will allow it to do so. This is particularly true of the organic sector. However, we believe that the Government has a role in analysing the organic supply chain for bottlenecks and imbalances and devising policy tools to help remedy these (paragraph 91).**

PROCEEDINGS OF THE COMMITTEE
RELATING TO THE REPORT

WEDNESDAY 17 JANUARY 2001

[AFTERNOON MEETING]

Members present:

Mr David Curry, in the Chair

Mr David Drew   Mr Mark Todd
Mr Michael Jack   Dr George Turner
Mr Austin Mitchell

The Committee deliberated.

Draft Report [Organic Farming], proposed by the Chairman, brought up and read.

Ordered, That the draft Report be read a second time, paragraph by paragraph.

Paragraphs 1 to 92 read and agreed to.

Resolved, That the Report be the Second Report of the Committee to the House.

Ordered, That the Chairman do make the Report to the House.

Ordered, That the provisions of Standing Order No. 134 (Select committees (reports)) be applied to the Report. –(*The Chairman.*)

Ordered, That the Appendices to the Minutes of Evidence taken before the Committee be reported to the House. –(*The Chairman.*)

[Adjourned till Wednesday 24 January at Ten o'clock.

LIST OF WITNESSES

LIST OF MEMORANDA
INCLUDED IN THE MINUTES OF EVIDENCE

Page

LIST OF APPENDICES
TO THE MINUTES OF EVIDENCE

LIST OF UNPRINTED MEMORANDA

Additional memoranda have been received from the following and have been reported to the House, but to save printing costs they have not been printed and copies have been placed in the House of Commons Library where they may be inspected by Members. Other copies are in the Record Office, House of Lords, and are available to the public for inspection. Requests for inspection should be addressed to the Record Office, House of Lords, London SW1 (Tel 020 7219 3074). Hours of inspection are from 9.30 am to 5.30 pm on Mondays to Fridays.

1. Mr John Whetman (F3) (Annexes)

2. The Organic Food and Farming Targets Bill Campaign (F15) (Annexes)

3. Horticulture Research International (F21) (Annex)

4. Mr Geoffrey Hollis (F25) (Annexes)

5. Consumers' Association (F28)

6. JG Quicke and Partners (F29) (Newsletter)

7. English Nature (F39) (Annexes 3 and 4)

8. National Office of Animal Health Ltd (F52)

9. National Office of Animal Health Ltd (F61)

10. Ms Anna Appelmelk (F65 [Annexes]) and (F66)

AGRICULTURE COMMITTEE REPORTS
IN THE CURRENT PARLIAMENT

Session 1997-98

FIRST REPORT, MAFF/Intervention Board Departmental Report 1997, HC 310, published on 16 December 1997.

SECOND REPORT, CAP Reform: Agenda 2000, HC 311-I, published on 25 February 1998.

THIRD REPORT, The UK Beef Industry, HC 474, published on 3 March 1998.

FOURTH REPORT, Food Safety, HC 331, published on 29 April 1998.

FIFTH REPORT, Vitamin B6, HC 753, published on 23 June 1998.

SIXTH REPORT, Flood and Coastal Defence, HC 707, published on 5 August 1998.

SEVENTH REPORT, Vitamin B6: The Government's Decision, HC 1083, published on 4 August 1998.

Session 1998-99

FIRST REPORT, MAFF/Intervention Board Departmental Report 1998 and the Comprehensive Spending Review, HC 125, published on 5 January 1999.

SECOND REPORT, CAP Reform: Rural Development, HC 61, published on 19 January 1999.

THIRD REPORT, The UK Pig Industry, HC 87, published on 1 February 1999.

FOURTH REPORT, UK Pig Industry: The Government's Response, HC 367, published on 6 April 1999.

FIFTH REPORT, Badgers and Bovine Tuberculosis, HC 233, published on 27 April 1999.

SIXTH REPORT, Genetically Modified Organisms, HC 427, published on 15 June 1999.

SEVENTH REPORT, Outcome of the CAP Reform Negotiations, HC 442, published on 29 June 1999.

EIGHTH REPORT, Sea Fishing, HC 141, published on 5 August 1999.

NINTH REPORT, MAFF/Intervention Board Departmental Report 1999, HC 852, published on 28 October 1999.

Session 1999-2000

FIRST REPORT, The Current Crisis in the Livestock Industry, HC 94, published on 14 December 1999.

SECOND REPORT, The Marketing of Milk, HC 36, published on 1 February 2000.

THIRD REPORT, The Segregation of Genetically Modified Foods, HC 71, published on 7 March 2000.

FOURTH REPORT, Environmental Regulation and Farming, HC 212, published on 17 March 2000.

FIFTH REPORT, The Government's Proposals for Organophosphate Sheep Dips, HC 425, published on 23 May 2000.

SIXTH REPORT, The Implications for UK Agriculture and EU Agricultural Policy of Trade Liberalisation and the WTO Round, HC 484, published 4 July 2000.

SEVENTH REPORT, Horticulture Research International, HC 484, published on 11 July 2000.

EIGHTH REPORT, Genetically Modified Organisms and Seed Segregation, HC 812 , published 3 August 2000.

NINTH REPORT, MAFF/IB Departmental Report 2000, HC 610, published 2 August 2000.

TENTH REPORT, Regional Service Centres, HC 509, published 1 August 2000.

Session 2000-2001

FIRST REPORT, Badgers and Bovine Tuberculosis: Follow-up, HC 92, published 10 January 2001.

ISBN 0-10-203901-1

Printed in the United Kingdom by The Stationery Office Limited
1/2001   584849   19585   CRC Supplied